2022 IMAPS Nordic Conference on Microelectronics Packaging (NordPac 2022)

Gothenburg, Sweden
12 – 14 June 2022

IEEE Catalog Number: CFP22M25-POD
ISBN: 978-1-6654-9177-8

Copyright © 2022, IMAPS Nordic
All Rights Reserved

*** *This is a print representation of what appears in the IEEE Digital Library. Some format issues inherent in the e-media version may also appear in this print version.*

IEEE Catalog Number: CFP22M25-POD
ISBN (Print-On-Demand):   978-1-6654-9177-8
ISBN (Online):   978-91-89711-39-6

Additional Copies of This Publication Are Available From:

Curran Associates, Inc
57 Morehouse Lane
Red Hook, NY 12571 USA
Phone: (845) 758-0400
Fax: (845) 758-2633
E-mail: curran@proceedings.com
Web: www.proceedings.com

2022 IMAPS Nordic Conference on Microelectronics Packaging (NordPac 2022)

Gothenburg, Sweden
12 – 14 June 2022

IEEE Catalog Number: CFP22M25-POD
ISBN: 978-1-6654-9177-8

Table of contents

Session M1B: Advanced packaging solutions

Low Inductive SiC Power Electronics Module with Flexible PCB Interconnections and 3D Printed Casing...1
Saeed Akbari, Rise Research Institutes of Sweden.

BiCMOS Integrated Temperature Sensor for Thermal Evaluation of Fan-out Wafer-level Packaging (FOWLP) including Hot Spot Analysis...8
Matthias Wietstruck, IHP - Leibniz Institut für innovative Mikroelektronik

Session T2A: Thermal management

Optimising thermal management of MEMS element with thermoelectric-cooler...12
Daniel Nilsen Wright, SINTEF Digital

High Aspect Ratio Through-Glass Vias As Heat Conductive Element...18
Kevin Kröhnert, Fraunhofer IZM

Reliability Characterization of Graphene Enhanced Thermal Interface Material for Electronics Cooling Applications...24
Markus Enmark, Chalmers University of Technology

Session T2B: Materials

Soft and Stretchable Wireless Sensor Patch with Digitally Printed Liquid Metal Alloy Interconnects...30
Jan Maslik, Uppsala University, Sweden

CHALMERS

2022 IMAPS Nordic Conference on Microelectronics Packaging (NordPac)

June 12-14, 2022
Göteborg, Sweden

NordPac 2022 Annual Microelectronics and Packaging Conference and Exhibition

June 12 – 14

Chalmers University of Technology

Gothenburg, Sweden

Reviewed Papers

CHALMERS

Exhibitors

Welcome

On behalf of IMAPS Nordic, IEEE EPS Nordic and the local organizing committee at Chalmers University of Technology, I would like to welcome you to NordPac 2022, the Nordic Microelectronics and Packaging conference. This is the fourth time that IMAPS Nordic Chapter and IEEE EPS Nordic Chapter jointly organise this event. The NordPac conference is a strong platform that brings together academics, as well as industry leaders, to discuss and debate state-of-the-art and future trends in microelectronics components, packaging, integration, and manufacturing technologies.

This year we are located at the campus of Chalmers University of Technology, Gothenburg. We have continued the tradition of offering free short courses to all registered participants in addition to offering a varied program and exciting exhibitors.

I hope you enjoy the conference and exhibition and value the networking possibilities during the coffee&tea breaks and lunches. In the conference dinner we are finally able to celebrate IMAPS Nordic 50 years anniversary.

Warmly welcome to experience the event in a beautiful Nordic summer,

Heidi Lundén

President of IMAPS Nordic.

IMAPS Nordic board:

Heidi Lundén	Schott Primoceler Oy	Finland
Daniel Nilsen Wright	SINTEF	Norway
Anders E. Petersen	Oticon	Denmark
Dag Andersson	RISE IVF AB	Sweden
Rajan Ambat	Technical University of Denmark	Denmark
Johan Liu	Chalmers University of Technology	Sweden
Terho Kutilainen	Oy Poltronic Ab	Finland
Paul Collander	Retired	Finland

Local organizing committee:

Johan Liu (Chalmers University of Technology)
Yifeng Fu (Chalmers University of Technology)
Susannah Carlsson (Chalmers University of Technology)
Gunnel Berggren (Chalmers University of Technology)
Markus Enmark (Chalmers University of Technology)
Athanasios Theodoridis (Chalmers University of Technology)
Zhen Li (Chalmers University of Technology)
Sihua Guo (Chalmers University of Technology)
Hafid Zehri

CHALMERS

Low Inductive SiC Power Electronics Module with Flexible PCB Interconnections and 3D Printed Casing

Saeed Akbari*, Konstantin Kostov, Klas Brinkfeldt, Mietek Bakowski, Dag Andersson
RISE Research Institutes of Sweden, Mölndal, Sweden
Email: saeed.akbari@ri.se

Abstract— *Silicon carbide* (SiC) power devices are steadily increasing their market share in various power electronics applications. However, they require low-inductive packaging in order to realize their full potential. In this research, low-inductive layouts for half-bridge power modules, using a direct bonded copper (DBC) substrate, that are suitable for SiC power devices, were designed and tested. To reduce the negative effects of the switching transients on the gate voltage, flexible printed circuit boards (PCBs) were used to interconnect the gate and source pins of the module with the corresponding pads of the power chips. In addition, conductive springs were used as low inductive, solder-free contacts for the module power terminals. The module casing and lid were produced using additive manufacturing, also known as 3D printing, to create a compact design. It is shown that the inductance of this module is significantly lower than the commercially available modules.

Keywords— Power electronics packaging, low inductive module, SiC devices, parasitic inductance.

I. INTRODUCTION

ALTHOUGH silicon semiconductors have been used as the main switching devices in power electronics for many years, they are rapidly approaching their structural limits [1], and gradually being replaced by *wide bandgap* (WBG) semiconductors. There are two WBG materials that are already commercially available: *Silicon Carbide* (SiC) and *Gallium Nitride* (GaN). SiC *metal–oxide semiconductor field-effect transistors* (MOSFETs) have already gained a significant market share in the medium and high-voltage power applications.

SiC devices offer many advantages over their silicon counterparts [2-3]. From efficiency point of view, the main advantages of SiC are lower on-state resistance and higher switching speed, which lead to lower conduction and switching losses, respectively. However, the relatively high parasitic inductance of conventional power modules leads to voltage overshoot and limits the switching speed. This and other limitations of the old packaging technologies are preventing the full potential of SiC to be realized. In order to utilize the high switching speed of these devices, novel packaging designs with low parasitic inductance are required [4-5].

This work describes a power module with SiC MOSFETs, which aims to minimize the parasitic inductance of planar power modules built with *direct bonded copper* (DBC) substrates. Flexible *printed circuit boards* (PCBs) were used to interconnect the module gate and source pins to the corresponding pads on the SiC dies. The flexible PCBs were designed to shield the gate signals from *electromagnetic*

interference (EMI) and minimize the internal parasitic inductances between the gate and source pins of the module and the corresponding pads on the SiC dies. The parasitic inductances of the module were calculated with ANSYS Q3D. The experimental results of the switching performance are also presented.

II. POWER MODULE DESIGN AND ASSEMBLY

This section presents the design and fabrication of the low inductive SiC power module. The manufacturing steps of the *half-bridge* (HB) module are explained in Figs. 1-2. The module substrate was a DBC, which is a copper-ceramic-copper laminate, with a total thickness of 1.23 mm. The ceramic layer of the DBC was *aluminium nitride* (AlN) with 0.63 mm thickness and thermal conductivity of 170 W/mK, which provides a low thermal resistance path for the heat generated by the SiC chips on the top side, to the base plate, attached on the bottom side of the DBC substrate.

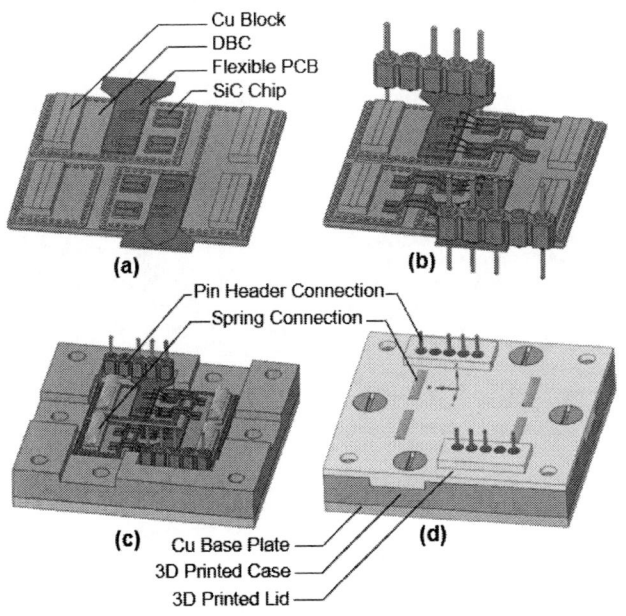

Fig. 1. Assembly steps of the SiC power module. a) Soldering the power devices, the flexible PCBs, and the copper blocks to the DBC substrate. b) Wire and ribbon-bonding the power devices, then soldering the pin headers to the flexible PCBs. c) Bonding the 3D printed casing to the copper base plate, then the module DBC substrate to the copper base plate using a thermal paste. d) Attaching the 3D printed lid to the module using four non-conductive screws.

Dimples were used along the edges on the top copper layer. They lower the interfacial stress, caused by the different *coefficients of thermal expansion* (CTE) of *copper* (Cu) and AlN. These dimples increase significantly the lifetime of the substrate, and therefore, the overall reliability of the power module [6-8].

At the first manufacturing step, the chips, the copper blocks, and the flexible PCBs were attached to the DBC using a lead-free soldering process with a lead-free paste (SAC305) in a reflow oven (Figs. 1a and 2a). The DBC and the flexible PCB both had *electroless nickel immersion gold* (ENIG) metallization. The power devices used here were 1200 V, 80 mΩ SiC MOSFETs from ON Semiconductor® with the configuration and dimensions shown in Fig. 3.

The gate and source pads of the chips were connected to the corresponding pads on the flexible PCBs with bondwires. Aluminium ribbons were used for the high-current connections between the source pads of the chips and the DBC (Figs. 1b and 2b). The ribbons were 1 mm wide and 0.1 mm thick, and the wire bonds were 0.125 mm thick. After ribbon and wire bonding, a silicone potting gel was applied on the device surface as well as the wire bonds and the ribbon bonds to cover them completely and protect them from environmental pollution and physical damage (Fig. 2c). This also improves the electrical safety and reliability by reducing the risk of surface flashover between different conductors.

As an alternative to the standard high inductance terminals, spring contacts were used to provide solder-free, surface-mount connection between the module and the external circuitry. The spring contacts had a wire thickness of 0.3 mm, and an outer diameter of 2.78 mm. As shown in Fig. 4, a curvature was created on the top surface of the copper blocks to increase the contact surface between the spring and the copper block and improve the electrical conductivity. These spring-type terminals have lower inductance compared to screw-type terminals because they are short and wide.

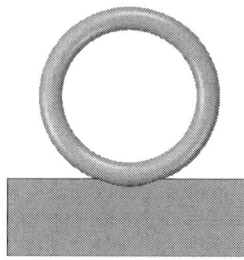

Fig. 4. The spring connection is placed on a copper block. To increase the contact surface, a curved slot was created on the copper surface where the spring sits.

Next, the 3D printed casing was bonded to the base plate using a moisture-cured super glue (Cyanoacrylate). The DBC module was then bonded to the base plate using a *thermal interface material* (TIM) adhesive to facilitate the heat dissipation from the DBC to the base plate. The TIM had a thermal conductivity of 0.7 W/mK, and was supplied by Electrolube. Finally, four non-conductive screws, made from a thermoplastic polyamide (Nylon 66), were used to attach the 3D printed lid to the 3D printed casing (Figs. 1d and 2d). The module is mounted to a heatsink with four metal M3 screws through the four corners of the package.

The power module lid and casing were manufactured by *stereolithography* (SLA), which is one of the common *additive manufacturing* (AM) techniques, also known as 3D printing. The package lid and casing, as shown in Figs. 1c-d, 2d and 5, consist of many intricate details to accommodate the module external connections. These connections include spring connections, pin header connections, copper blocks, and screws. These features are challenging to manufacture using the convectional moulding methods. AM is desired for rapid prototyping and testing different designs. However, conventional methods are preferred for mass production.

SLA is an ideal solution to produce the module casing and lid, because it is able to rapidly create highly accurate complex objects with smooth surfaces. In this method, the desired 3D model is created in a computer aided design (CAD) software. The CAD file is then converted into a language, or file format, that the 3D printing apparatus can understand. Standard Tessellation Language (STL) is the file format that is most commonly employed by the 3D printing machines. The STL file provides the machine with the information it requires to print the 3D object. During the printing process, a transparent photocurable polymer is exposed to a UV laser source, and hardened in a layer-by-layer manner. The printing process was

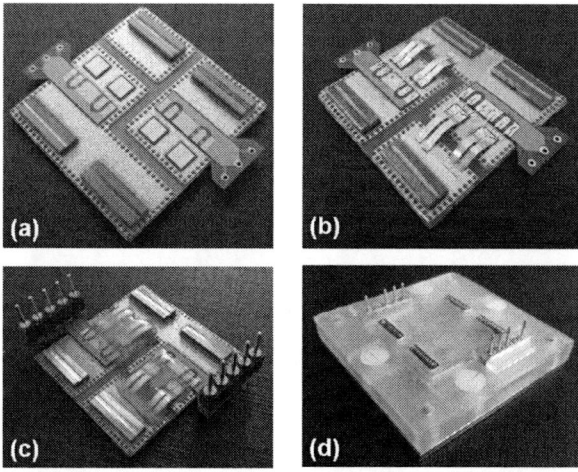

Fig. 2. The power module produced based on the steps described in Fig. 1.

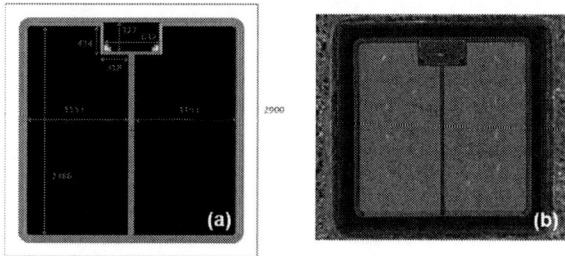

Fig. 3. a) Layout and dimensions of SiC MOSFET (1200 V, 80 mΩ). Dimensions in μm. The black regions are *photosensitive polyimide* (PSPI) openings. The green regions are pad metal covered with PSPI. b) The surface of the chip after soldering to the DBC.

done using a Formlabs machine.

In this work, AM was used to produce the module non-functional parts, i.e. the casing and the lid. This technology has recently been developed to fabricate functional parts such as electronics circuits. Ideally, it enables simultaneous deposition of dielectric and conductive layers with complicated 3D configurations to fabricate electronics assemblies with embedded devices and interconnections. To this end, the most promising AM techniques with commercially available electronics printers include inkjet [9-10], aerosol [11-12], and *laser induced forward transfer* (LIFT) [13]. While conventional electronics manufacturing includes many steps, printed electronics includes only one single step, regardless of the design complexity.

The flexible PCBs are soldered to the DBC and provide electric connectivity between the module gate/source pins and the corresponding MOSFET pads. The layout of the flexible PCBs is shown in Fig. 6. The first (top) layer has sufficiently large pads for the gate bondwires. The rest of the top layer is a copper pour connected to the source pin of the module, and provides enough space for the two bondwires connecting the two sources of each MOSFET. The second layer contains the traces connecting the gate pins of the module with the gate pads on the top layer.

To have individual control of the chips there are two gate pins - one for each MOSFET on the upper and lower switch position of the *half bridge* (HB). The gate traces were made to have the same length but can also be designed to have the same inductance. In most HB modules one gate pin per switch position is sufficient. The third layer is a copper pour connected to the source and to the top layer. Together, the copper pours on the first and third layers form a shield for the gate signals inside the module. Only the gate bondwires are exposed to the switching electromagnetic interference.

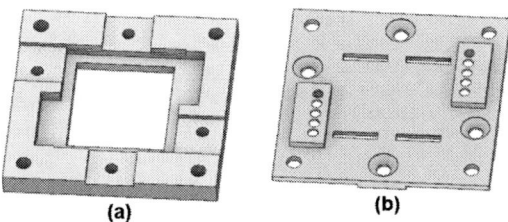

Fig. 5. The power module casing (a) and lid (b) produced by SLA 3D printing from an acrylate-based polymer.

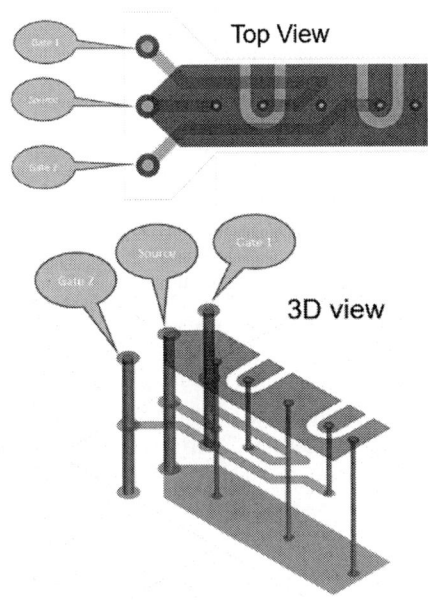

Fig. 6. The layout of the flexible PCB.

III. PARASITIC INDUCTANCE OF THE MODULE

Two layouts (Fig. 7) were considered in this work, referred to as the L-shaped and the U-shaped design. Figs. 1-2 describe the manufacturing steps for the L-shaped design. The production steps explained in Section II were identical for both L-shaped and U-shaped designs.

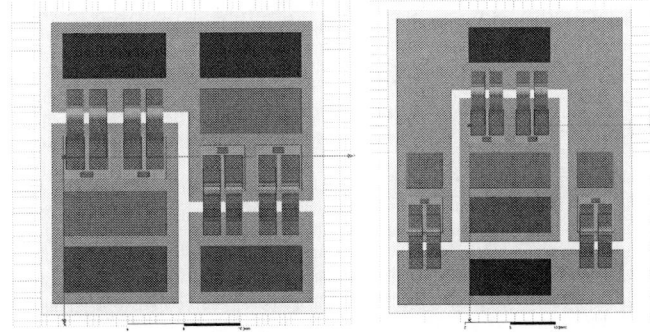

Fig. 7. L-shaped and U-shaped layouts. The green regions on the top of the DBC copper (apart from the SiC MOSFETs) mark the areas for attaching the flexible PCB. The red region shows the area for the copper block of the DC plus terminal, the blue one the DC minus terminal, and the dark brown region(s) the phase terminal(s).

As mentioned earlier, the parasitic inductances of the module must be as low as possible to allow for fast switching, which would minimize the switching losses. The parasitic inductances of these designs were calculated with ANSYS Q3D Extractor. The stray inductance of the power module, which is usually given in its datasheet, is defined as the parasitic inductance between the DC plus and minus terminals. For the L-shaped and U-shaped layouts these were respectively 11.8 nH and 11.2 nH (Table I). The U-shaped layout has smaller stray inductance because the DC plus terminal is closer to the phase terminal, and since the current loop flows via the phase copper region, it forms a relatively smaller loop. This can be further reduced by bringing the DC plus terminal closer to the MOSFETs in the middle, but in the present design, the idea was to use a single flexible PCB.

Table I: Parasitic inductances for the two planar layouts.

Parasitic inductance		
Between	L-shaped layout	U-shaped layout
(Plus)-(Minus)	11.83 nH	11.16 nH
(Plus)-(Phase)	8.72 nH	8.61 nH
(Minus)-(Phase)	5.94 nH	3.66 nH

In normal operation, the current switches between the phase terminal and one of the DC terminals. It should not flow between DC plus and DC minus. Therefore, it would make sense to determine the parasitic inductances between DC plus and phase terminals, as well as between DC minus and phase terminals. The parasitic inductance between DC plus and phase terminals for the L- and U-shaped layouts were 5.94 nH and 3.7 nH, respectively, and those between the phase and DC minus terminals were 8.7 nH and 8.6 nH respectively. This shows that although the U-shaped layout has a lower stray inductance, there is a larger difference between the parasitic inductances of the upper and lower switch positions of the HB, which may lead to slower switching of the lower MOSFETs. Although the L-shaped layout has larger stray inductance, it should provide more equal switching speed between the upper and lower MOSFETs.

Paralleling of more semiconductor chips is possible for both layouts. This would require increasing only the horizontal dimension of the modules (the other two dimensions remain the same), and the current density in the copper regions of the DBC will remain the same regardless of the number of paralleled chips.

IV. SWITCHING PERFORMANCE

Only the switching performance of the module with the L-shaped layout was tested. Fig. 8 shows the switching test setup to drive the primary windings in a DC-DC converter. Fig. 9 shows the continuous switching test. From electrical point of view (clearance and rating of the chips inside), the modules are rated for 1.2 kV, but due to insufficient cooling and limitations of the DC power supply, the DC voltage in this test was only 100 V. Rogowski coil was used to measure the phase current.

Fig. 8. Switching test setup with two modules (under the PCB and on top of the heatsink). The DC link capacitors are visible on top.

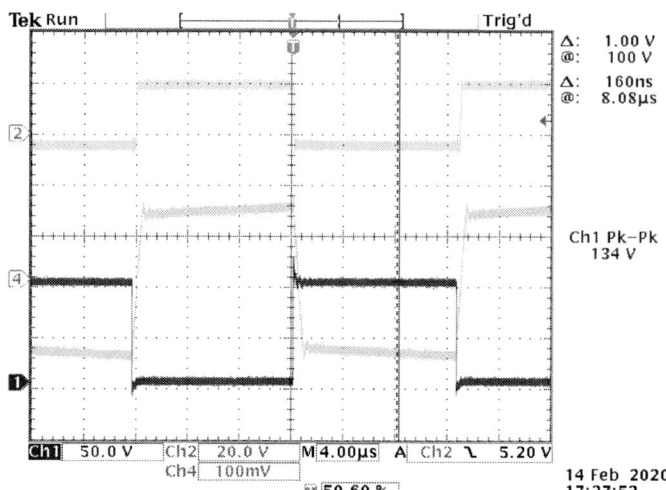

Fig. 9. Overview of the continuous switching. Ch1 is drain voltage over the lower MOSFETs, Ch2 is the gate voltage for the lower MOSFETs and Ch4 is the output current (10A/div).

The zoom in of the turn-off in Fig. 9 is shown in Fig. 10, which shows that the voltage rise-time is less than 10ns.

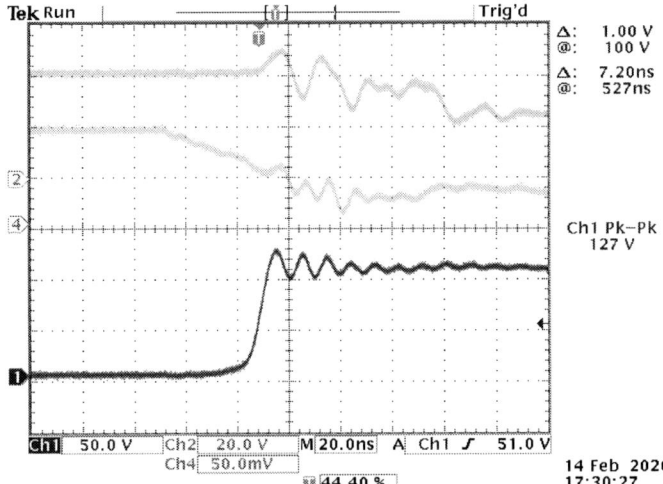

Fig. 10. A zoom in of the turn-off in Fig. 9 at 15 A phase current. Ch1 is the drain-voltage, Ch2 is the gate voltage and Ch4 is the output current (5 A/div).

Direct measurement of the parasitic inductances of the module is almost impossible, but the ringing in the voltage waveforms can be used to estimate the parasitic inductances indirectly. Because of the limited bandwidth of the current probes, the oscillations in the current waveform cannot be used for this purpose. Fig. 11 shows the turn-off waveforms at 20 A phase current, which was the limit of the supply feeding the DC link. At this current the oscillations in the turn-off voltage are larger, which makes it easier to determine their frequency.

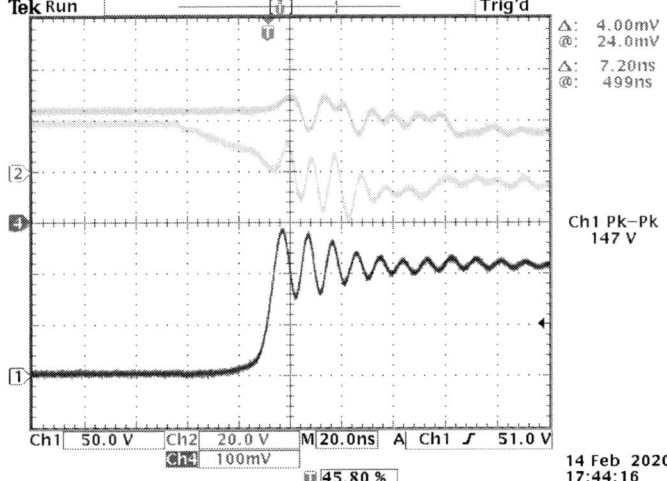

Fig. 11. Turn-off of the lower MOSFETs at 20 A phase current. Ch1 is the drain-voltage, Ch2 is the gate voltage and Ch4 is the output current (10 A/div).

The voltage ringing at turn-off in Fig. 11 settles at around 100 MHz. Knowing that the output capacitance of the MOSFETs at 100 V is 400 pF (200 pF each) the parasitic inductance can be calculated to be 6.33 nH, but this includes the parasitic inductance of the trace between the Minus terminal of the module and the DC link capacitor. This confirms the Minus-to-Phase parasitic inductance calculated by ANSYS Q3D. Unfortunately, the turn-off voltage of the upper MOSFETs was not measured. It would have allowed the calculation of the Plus-to-Phase parasitic inductance.

Fig. 12 shows the passive turn-on of the lower MOSFETs, i.e. the turn-on at negative current when the internal body diodes of the MOSFETs are conducting (the gate turn-on signal comes more than 200 ns after the MOSFETs have turned-on). The voltage waveform shows that the voltage turn-on time is very fast (around 10 ns).

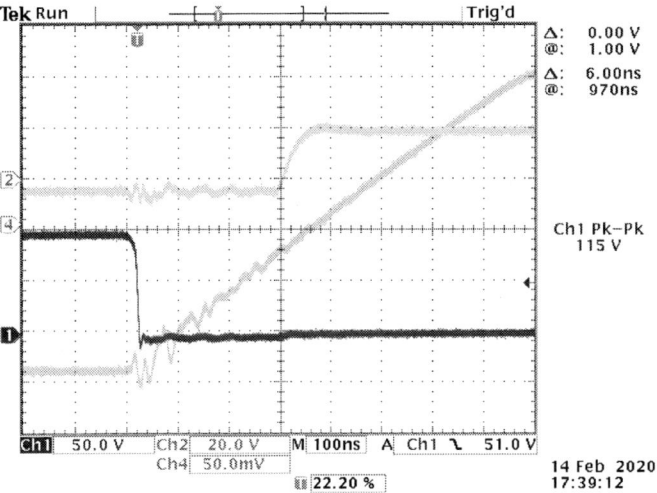

Fig. 12. Passive turn-on at 15 A phase current. Ch1 is the drain-voltage, Ch2 is the gate voltage and Ch4 is the output current (5 A/div).

In addition to the continuous switching tests, the double-pulse test was also performed (Figs. 13-15), but for that we could increase the DC supply voltage to 200 V. An inductor of 0.750 mH was connected between the phase terminal and the positive DC bus line. Then the *device under test* (DUT) is the lower MOSFET. As seen in Fig. 13, the turn-off of the DUT occurs when the inductor current reaches 40 A. The turn-off of the DUT is the start of the freewheeling period during which the inductor current flows through the internal body diodes of the upper MOSFETs. After 20 µs the DUT is turned-on again. During the freewheeling period, there is a slight drop in the inductor current due to the forward voltage drop of the freewheeling diodes and due to the energy losses around the freewheeling loop. Therefore, the subsequent turn-on occurs at slightly less than 40 A.

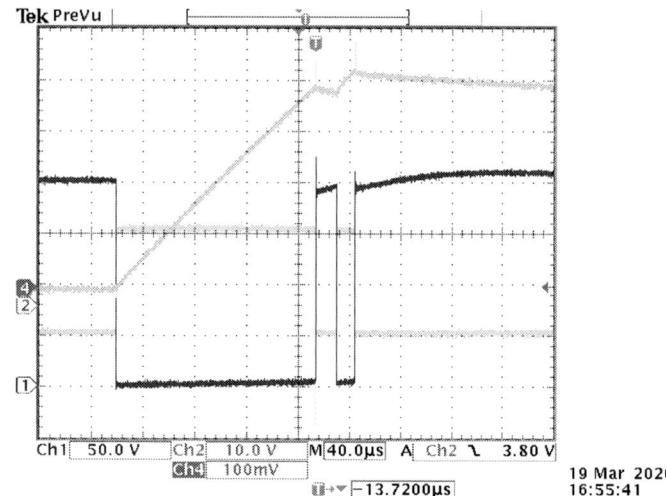

Fig. 13. Overview of the double pulse test waveforms. Ch1 is drain voltage, Ch2 is gate voltage, and Ch4 is inductor current.

Fig. 14 is a zoom in the turn-on waveforms from Fig. 13. It shows that the current rise time is about 25 ns.

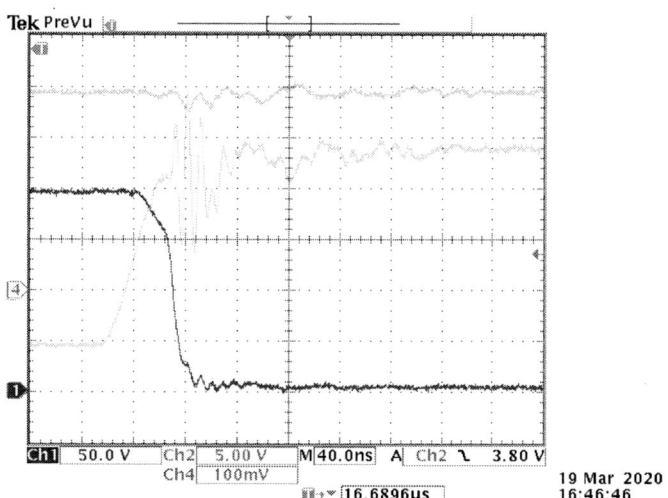

Fig. 14. Turn-on waveforms at 40 A inductor current.

Unfortunately, the zoom in the turn-off waveforms from the double-pulse test were too noisy and it is difficult to give precise number for the current fall time.

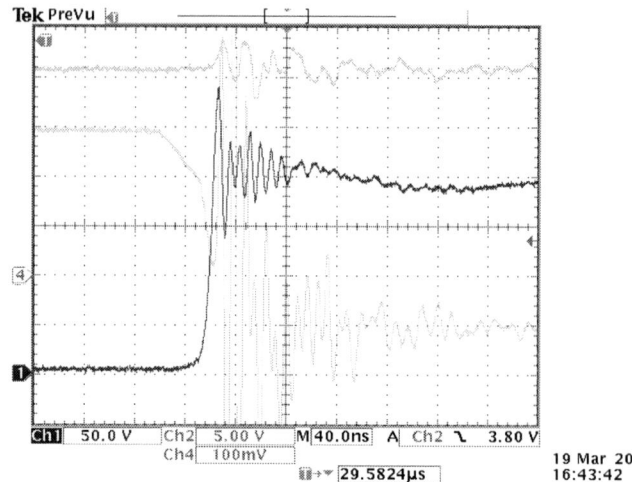

Fig. 15. Turn-off waveforms at 40 A inductor current

V. CONCLUSIONS

In order to reduce the switching losses and oscillations, the parasitic inductance of the module must be minimized. This paper presents two layouts suitable for planar SiC power modules, using DBC, with multiple power devices in parallel. Flexible PCBs were used internally to connect the gate-source pins of the module with the corresponding pads on the power semiconductor chips, and to shield the gate signals, which protects the latter form interference and allows synchronize the gate signals for the paralleled devices. 3D printing was used to manufacture the case and the lid of the module.

It was shown that the proposed planar layouts have very small parasitic inductances, which allows for very fast switching. Further reduction of the parasitic inductance of planar power modules on DBC substrates is possible by developing new types of low-inductive DC-link terminals, or by replacing the ribbons on the top side of the MOSFET chips with copper clips or other types of conductive substrate.

In addition to minimizing the parasitic inductance, more research is needed to make the modules more symmetrical, i.e. to make the parasitic inductances from Phase to the two DC terminals equal.

ACKNOWLEDGMENT

This work was performed under the project Low-Inductive SiC Module (LISM) that received funding from the Swedish energy agency Energimyndigheten with the grant agreement No 44163.

REFERENCES

[1] Millan J, Godignon P, Perpiñà X, Pérez-Tomás A, Rebollo J. A survey of wide bandgap power semiconductor devices. IEEE transactions on Power Electronics. 2013 Jun 14;29(5):2155-63.

[2] Rabkowski J, Peftitsis D, Nee HP. Silicon carbide power transistors: A new era in power electronics is initiated. IEEE Industrial Electronics Magazine. 2012 Jun 15;6(2):17-26.

[3] Palmour JW, Cheng L, Pala V, Brunt EV, Lichtenwalner DJ, Wang GY, Richmond J, O'Loughlin M, Ryu S, Allen ST, Burk AA. Silicon carbide power MOSFETs: Breakthrough performance from 900 V up to 15 kV.

In2014 IEEE 26th International Symposium on Power Semiconductor Devices & IC's (ISPSD) 2014 Jun 15 (pp. 79-82). IEEE.

[4] DiMarino C, Mouawad B, Johnson CM, Wang M, Tan YS, Lu GQ, Boroyevich D, Burgos R. Design and experimental validation of a wire-bond-less 10-kv sic mosfet power module. IEEE Journal of Emerging and Selected Topics in Power Electronics. 2019 Sep 27;8(1):381-94.

[5] Gurpinar E, Chowdhury S, Ozpineci B, Fan W. Graphite-embedded high-performance insulated metal substrate for wide-bandgap power modules. IEEE Transactions on Power Electronics. 2020 Jun 16;36(1):114-28.

[6] Han L, Liang L, Chen D, Zhao Z, Luo F, Kang Y. Modeling and analysis of mesh pattern influences on DBC thermal cycling reliability. Microelectronics Reliability. 2020 Jul 1;110:113645.

[7] Shi Y, He H, Zhu W. Finite Element Analysis on Fracture Mechanics of Al_2O_3, AlN and Si_3N_4-based Substrates under Thermal Shock Condition. In2020 21st International Conference on Electronic Packaging Technology (ICEPT) 2020 Aug 12 (pp. 1-6). IEEE.

[8] Gaiser P, Klingler M, Wilde J. Fracture mechanical modeling for the stress analysis of DBC ceramics. In2015 16th International Conference on Thermal, Mechanical and Multi-Physics Simulation and Experiments in Microelectronics and Microsystems 2015 Apr 19 (pp. 1-6). IEEE.

[9] Schleicher M, Storz V, Kujath M. Disruptive Approach of Additive Manufactured Electronics (AME). In PCIM Europe digital days 2021; International Exhibition and Conference for Power Electronics, Intelligent Motion, Renewable Energy and Energy Management 2021 May 3 (pp. 1-8). VDE.

[10] Kimionis J, Isakov M, Koh BS, Georgiadis A, Tentzeris MM. 3D-printed origami packaging with inkjet-printed antennas for RF harvesting sensors. IEEE Transactions on Microwave Theory and Techniques. 2015 Nov 11;63(12):4521-32.

[11] Sarobol P, Cook A, Clem PG, Keicher D, Hirschfeld D, Hall AC, Bell NS. Additive manufacturing of hybrid circuits. Annual Review of Materials Research. 2016 Jul 1;46:41-62.

[12] Hines DR, Gu Y, Martin AA, Li P, Fleischer J, Clough-Paez A, Stackhouse G, Dasgupta A, Das S. Considerations of aerosol-jet printing for the fabrication of printed hybrid electronic circuits. Additive Manufacturing. 2021 Nov 1;47:102325.

[13] Zacharatos F, Makrygianni M, Zergioti I. Laser-Induced Forward Transfer (LIFT) technique as an alternative for assembly and packaging of electronic components. IEEE Journal of Selected Topics in Quantum Electronics. 2021 May 27;27(6):1-8.

BiCMOS Integrated Temperature Sensor for Thermal Evaluation of Fan-out Wafer-level Packaging (FOWLP) including Hot Spot Analysis

Matthias Wietstruck*, Thomas Mausolf*, Jens Lehmann*, Zhibo Cao*, Thanh Duy Nguyen[†], Markus Wöhrmann[†], Tanja Braun[†]

*IHP – Leibniz Institut für innovative Mikroelektronik,
Im Technologiepark 25, 15230 Frankfurt (Oder), Germany
Email: wietstruck@ihp-microelectronics.com
[†] Fraunhofer-Institut für Zuverlässigkeit und Mikrointegration (IZM),
Gustav-Meyer-Allee 25, 13355 Berlin, Germany

Abstract—A SiGe BiCMOS thermal chip for thermal package characterization is developed and demonstrated together with a chip-first, face-down FOWLP technology. The dedicated BiCMOS thermal chip enables a spatially resolved heat generation and temperature characterization to analyze the effects of uniform heat generation as well as hot spots within integrated circuits and package technologies. The BiCMOS thermal chip can be applied for various wafer-level package technology providing an in-depth analysis of the thermal management.

Keywords—SiGe BiCMOS; Wafer-level Packaging, FOWLP; Thermal Management; Hot Spots

I. INTRODUCTION

Fan-out Wafer-level Packaging (FOWLP) is one of the key enabling technologies for wafer-level packaging (WLP) and 2.5/3D heterogeneous integration. A wide range of wafer-level packaging solutions e.g. substrate-less FOWLP, silicon/glass interposer or chip embedding have been established enabling an electrical redistribution of integrated circuits. Especially for millimeter wave applications, the combination of high performance SiGe BiCMOS together with FOWLP is highly desired due to their superior RF performance, high level of integration and low costs [1].

One of the major challenges for WLP and FOWLP is the thermal management and the related chip/package reliability [2]. With increasing power density and miniaturization, the thermal management becomes increasingly challenging. Although FOWLP enables a heat dissipation over the backside of the silicon chip, the further heat management becomes crucial affecting the overall thermal resistance R_{th}. Extensive FEM simulations are highly desired but they are mostly simplified without considering important thermal effects. As an example, a distributed heat source rather than local heat sources are mainly considered although hot spots can lead to a significant higher R_{th} due to the spreading resistance effect [3]. In addition, the thermal interface resistances, which occur e.g. in between the thermal interface material (TIM) and the silicon chip are often neglected. Those simplifications lead to an underestimation of the overall package R_{th} thus higher chip temperatures occur which limits the reliability and lead to performance variations. To overcome

those limitations the accurate determination of the thermal performance of CMOS/BiCMOS chips with FOWLP is mandatory. Dedicated thermal chips for thermal characterization are available but they can suffer from technology variations compared to the application-specific integrated circuit technologies (e.g. layer stack-up, exact position of heat source, surface roughness, etc.) [4]. To overcome those limitations, dedicated CMOS/BiCMOS thermal chips for the chip-package thermal evaluation are highly desired.

For the accurate determination of the thermal performance of WLP/FOWLP technologies, a SiGe BiCMOS thermal chip is developed. The potential of the BiCMOS thermal chip for thermal/thermo-electrical package characterization is demonstrated by a chip-first, face-down FOWLP technology. The standalone thermal chip including several heaters and temperature sensors as well as the thermal chip in a FOWLP is characterized. Different power dissipation concepts are analyzed and the thermal distribution inside the chip is analyzed with a 2D thermal mapping showing the importance of local hot spots. The BiCMOS thermal IC is demonstrated enabling the accurate and in-depth thermal characterization of SiGe BiCMOS WLP, FOWLP and 2.5/3D heterogeneous integration.

II. BICMOS EMBEDDED THERMAL CHIP

A. BiCMOS Thermal Chip Design and Fabrication

For the accurate determination of the thermal and thermo-electrical performance of substrate integration and wafer-level packaging technologies, a SiGe BiCMOS thermal chip is developed. The thermal chip comprises 5 heaters based on polysilicon resistors and 16 temperature sensors using the collector-emitter junction of available heterojunction bipolar transistors. Each heater has a size of ~380x280 μm^2 whereas the temperature sensor have comparable small dimensions of ~50x150 μm^2. The heaters and temperature sensors are distributed over the chip enabling a spatially resolved heat and hot spot generation as well as a 2D thermal mapping on the chip. The fabricated BiCMOS chip and a schematic are illustrated in Fig 1. The overall chip size is 2.5x2.0 mm^2 and it is fabricated using IHPs 130 nm SiGe BiCMOS technology SG13S [5].

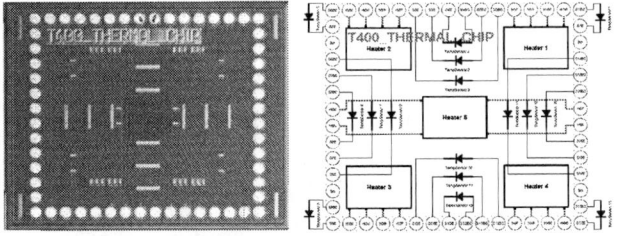

Fig. 1. Microscope image of the developed BiCMOS temperature chip and the related schematic showing the distributed heaters and sensors.

B. BiCMOS Thermal Chip Characterization

The BiCMOS embedded thermal chip is initially characterized to determine the functionality and performance of the heaters and temperature sensors. The individual components are characterized at wafer-level to extract the maximum limits and evaluate the in-die performance and uniformity. The measurement results for both the heaters and the temperature sensor are shown in Fig. 2 and Fig. 3. In case of the heater, a DC voltage up to 15 V is applied in a kelvin-type configuration and the dissipated power is extracted. A maximum power of >10 W/heater can be generated if the wafer is directly placed on a chuck. The non-linearity is caused by the negative temperature coefficient of resistance (TCR), which leads to a drop of the resistance for higher temperatures. The high available output power of the individual heater is mainly related to the characterization at wafer-level with a chuck which provides a huge thermal capacity thus the generated heat is directly distributed within the silicon substrate and the metal chuck limiting the temperature increase. A minimum variation between the different heaters can be observed.

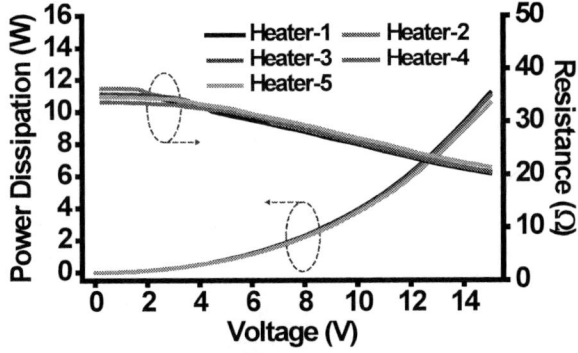

Fig. 2. Power dissipation & resistance vs. DC voltage of 5 different heaters.

Beside the heaters, the temperature sensors are characterized in the temperature range from 25-125 °C based on I-V measurements on a temperature-controlled chuck. The temperature sensor I-V characteristics are summarized in Fig. 2. The expected temperature-dependent diode characteristic can be clearly observed. Based on a fixed current of 1 μA, which ensures a negligible self-heating of the sensor itself, a linear behavior with a sensitivity of ~1.85 mV/K can be extracted enabling an accurate extraction of the chip temperature.

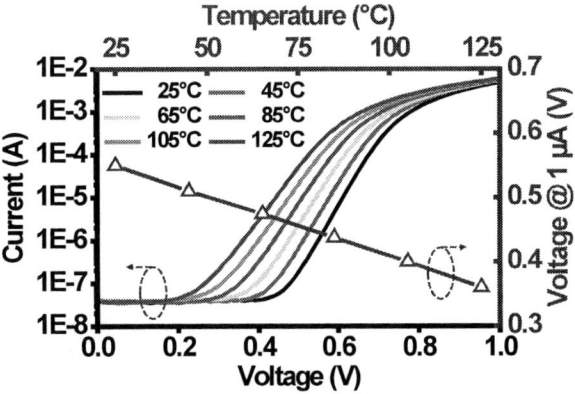

Fig. 3. Temperature sensor current-voltage characaristics and extracted diode voltage for a fixed current of 1 μA vs. chip temperature from 25-125°C.

III. BiCMOS EMBEDDED THERMAL CHIP WITH FOWLP

A. Fan-Out Wafer-level Packaging Technology

The potential of the BiCMOS thermal chip for thermal/thermo-electrical characterization of different substrate and wafer-level packaging technologies is demonstrated by a chip-first, face-down 200 mm FOWLP technology [6]. The thermal chip is packaged using a single redistribution layer for fanning out the signals of the individual heaters and temperature sensors. The overall package size is 8x8 mm². Solder balls with 240 μm are applied for the package assembly on a printed circuit board. A microscope image of the package is shown in Fig. 4. After finalizing the FOWLP process, the mold material on backside is removed by grinding to release the backside of the Si chip providing an ideal thermal path for heat dissipation.

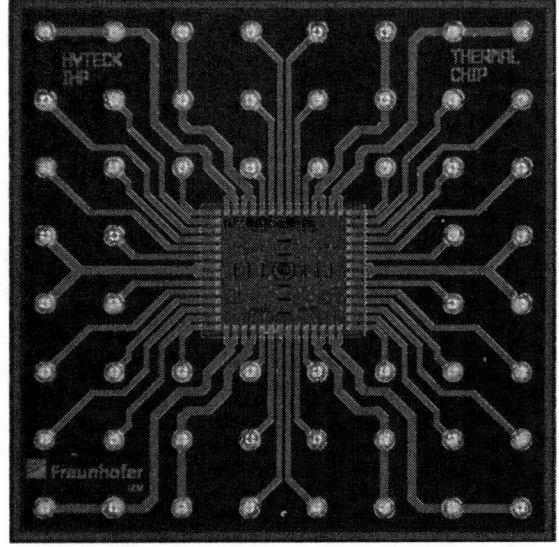

Fig. 4. Microscope image of showing the BiCMOS thermal chip in the center and the RDL for fanning out.

B. Thermal FEM Simulations

The BiCMOS embedded thermal chip with 5 distributed heaters and 16 temperature sensors enable complex thermal loads with either a quasi-uniform heat generation if all heaters are operating at the same time with the same thermal load or a hot spot heat generation if only individual heaters are active. FEM simulations are applied to investigate the thermal chip together with a FOWLP. The FEM model is shown in Fig. 5 (middle), which comprises the thermal chip in a FOWLP simulated with a cold plate with a fixed temperature of 25°C at the bottom. The epoxy mold compound (EMC) with a thermal conductivity λ=0.5 W/(m*K) as well as the silicon chip with λ=150 W/(m*K) have a thickness of 200 μm. A TIM with a thickness of 100 μm and λ=5 W/(m*K) provide the thermal interface to the Cu lid which has a thickness of 500 μm and λ=400 W/(m*K). At the bottom, an ambient temperature of 25°C is considered as boundary condition. Five different heat generation configurations with an overall defined thermal power of 10 W are evaluated: 10 W applied to the center heater, 10 W applied to the edge heater, in total 10 W applied to 3 diagonal heater (each heater 3.33 W) and in total 10 W uniformly applied to the 5 heaters (each heater 2 W). The simulated temperature distribution is shown in Fig. 5.

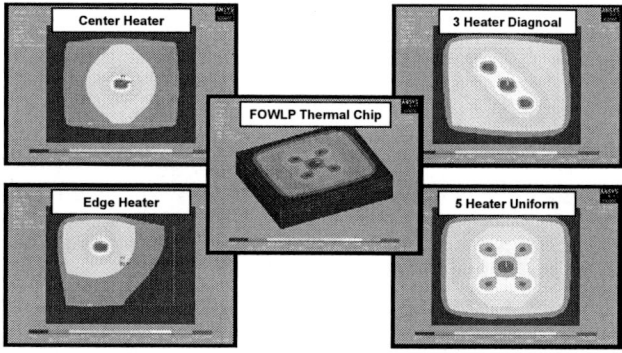

Fig. 5. FEM simulation results of the FOWLP thermal chip with different thermal loads.

The maximum temperature is extracted for the 4 different cases and the results are summarized in Fig. 6. A significant difference of the maximum chip temperature can be observed. If the center heater is active as an isolated heat source, the simulated temperature T_{max} is ~182.1°C whereas in case of an edge heater, T_{max} is ~193.9°C. This is caused by the higher thermal spreading resistance at the edge of the chip due to the different ratio of silicon and EMC compared to the center heater with a limited influence of the EMC. If the overall dissipated power of 10 W is applied to the 3 diagonal or 5 uniform heaters, the simulated temperature is significantly reduced to values of ~105.8 or 90.7°C, respectively. In case of the distributed heat generation, a significant lower temperature is achieved. This implies, that hot spots with a local rather than a global heat generation can lead to significant higher local temperatures. This can be extremely critical for a reliable chip operation as the maximum temperature is typically limited to values of 125°C for silicon-based CMOS/BiCMOS technologies.

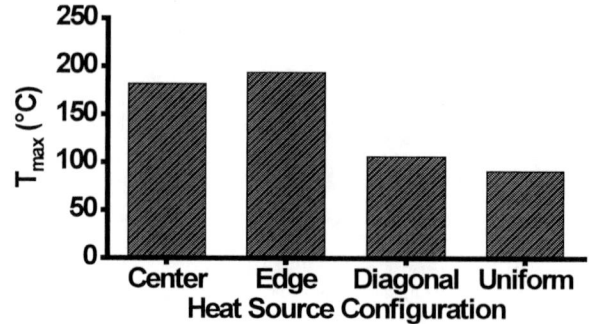

Fig. 6. Simulated maximum temperature T_{max} for the thermal chip with FOWLP considering 4 different heat source configurations.

C. FOWLP Printed Circuit Board Assembly

A test board is designed and fabricated enabling the electrical connection to the individual heaters and temperature sensors. The FOWLP including the SiGe BiCMOS thermal chip is flip chipped on the test board and a reflow soldering process is performed. A microscope image and an X-ray image of the assembled FOWLP thermal chip are shown in Fig. 7.

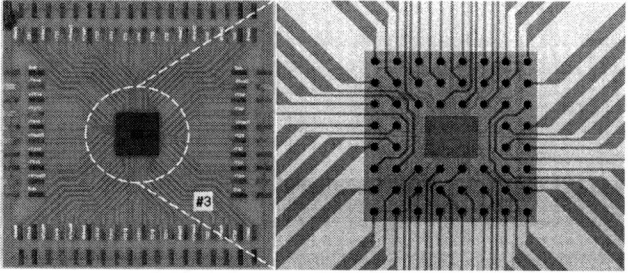

Fig. 7. Assembled FOWLP on PCB with additional X-ray analysis showing the accurate assembly step of the FOWLP to the PCB test board.

For the further thermal characterization of the thermal chip together with the FOWLP, an additional TIM tape with a thickness of 127 μm and λ_{TIM}=0.37 W/(m*K) together with a commercial available heat sink with an R_{th} of 7.89 K/W (in case of natural convection) is applied. The final PCB test board is shown in Fig. 8 together with the final setup for further thermo-electrical characterization.

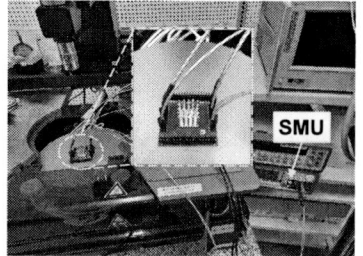

Fig. 8. Photograph of assembled FOWLP on PCB test board together with a heat sink (left) and test setup for thermo-electrical characterization (right).

D. Thermo-Electrical Characterization

The thermo-electrical characterization of the assembled SiGe BiCMOS chip with FOWLP is initiated with a calibration step of the different temperature sensors. Each sensor is characterized at a fixed temperature of 25/65/125°C and the final calibration parameter are extracted to be able to directly detect the temperature based on the measured diode voltage for a fixed current of 1 μA. Although a good uniformity of the sensor performance is achieved, a calibration is desired to improve the accuracy of the temperature measurements. As shown before various heat generation configurations can be realize enabling an in-depth characterization of the SiGe BiCMOS chip in combination with the FOWLP and the applied thermal management. Two different scenarios with a fixed power dissipation of 1.22 W at the center chip (scenario-1) and the same power dissipation distributed over the 5 different heaters (scenario-2) are presented. Scenario-1 can be considered as a hot spot analysis whereas scenario-2 is considered as uniform heat distribution. The total power is ramped and the temperature sensor signals are measured. As an example the measured temperature diode voltages and the extracted sensor temperatures for the scenario-1 for 5 selected temperature sensors are summarized in Fig. 9. The power ramp is limited to 1.22 W because a maximum temperature of ~125°C is present.

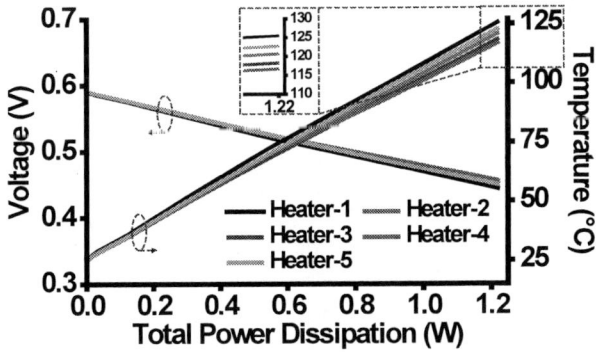

Fig. 9. Measuremet results of different temperature sensors for the scenario-1 with a maximum power dissipation of 1.22 W at the center heater.

A final illustration of the 2D temperature mapping is shown in Fig. 10. Scenario-1 shows a maximum temperature ranging from 125.5°C near the center heater down to 116.7°C at the edge of the Si chip. Scenario-2 shows a maximum temperature of only 113.1°C in the center and 111.6°C at the edge of the Si chip.

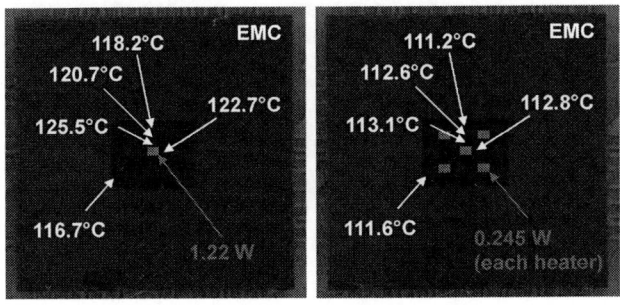

Fig. 10. Measuremet results of different temperature sensors for the scenario-1 (left) and scenario-2 (right) with an overall power dissipation of 1.22 W.

As expected a significant difference of the maximum temperature and the temperature variation on the chip can be observed if different thermal load scenarios are applied. In case of a hot spot thermal load with only one active heater, a higher maximum temperature and a higher temperature gradient over the chip can be clearly observed.

IV. CONCLUSION

For the accurate thermal performance characterization of WLP/FOWLP technologies, a SiGe BiCMOS thermal chip is developed. The BiCMOS thermal chip enables the thermal/thermo-electrical package characterization which is demonstrated by a chip-first, face-down FOWLP technology. The standalone thermal chip including several heaters and temperature sensors as well as the thermal chip in a FOWLP is characterized. Different power dissipation concepts are evaluated and the thermal distribution inside the chip is analyzed. Beside the straightforward thermal characterization of different wafer-level packaging technologies, the thermal chip can be also applied for the analysis and optimization of dedicated assembly and thermal management process steps as well as failure analysis and reliability tests. In summary the BiCMOS embedded thermal IC is demonstrated enabling the accurate and in-depth thermal characterization of SiGe BiCMOS WLP, FOWLP and heterogeneous integration.

ACKNOWLEDGMENT

The authors wants to thank IHP and IZM clean room team for excellent fabrication. This work was supported by the Federal Ministry of Education and Research (BMBF, Germany) as part of the project HyTeck "Hybridintegrationsplattform für zuverlässige Hochfrequenz-Schaltkreise" under Grant 16ES0713.

REFERENCES

[1] P. Chevalier et al., "SiGe BiCMOS Current Status and Future Trends in Europe," 2018 IEEE BiCMOS and Compound Semiconductor Integrated Circuits and Technology Symposium (BCICTS), 2018, pp. 64-71.

[2] T. Zhou and A. Samoilov, "Thermal management for wafer level packaging (WLP)," 2014 IEEE 64th Electronic Components and Technology Conference (ECTC), 2014, pp. 1679-1684.

[3] K. Chen, I. Hsu and C. Lee, "Chip-package-PCB thermal co-design for hot spot analysis in SoC," 2012 IEEE Electrical Design of Advanced Packaging and Systems Symposium (EDAPS), 2012, pp. 215-218.

[4] https://nanotest.eu/ttc.

[5] M. Kaynak et al., "0.13-μm SiGe BiCMOS technology with More-than-Moore modules," 2017 IEEE Bipolar/BiCMOS Circuits and Technology Meeting (BCTM), 2017, pp. 62-65.

[6] T. Braun et al., "Fan-out wafer level packaging for 5G and mm-Wave applications," 2018 International Conference on Electronics Packaging and iMAPS All Asia Conference (ICEP-IAAC), 2018, pp. 247-251.

Optimising thermal management of MEMS element with thermoelectric-cooler

Daniel Nilsen Wright*, Alessandro Liberale*, Andreas Vogl*, Niels Aakvaag*, Guillaume Savelli[†], Fabien Filhol[‡] and Romain Hodot[‡]

*SINTEF AS, Forskningsveien 1, 0373 Oslo, Norway
Email: Daniel.nilsen.wright@sintef.no
[†]CEA Liten, Grenoble, France
Email: guillaume.savelli@cea.fr
[‡]Thales AVS France SAS, Valence, France

Abstract— **The accuracy of MEMS sensing devices can be greatly increased by controlling the operating temperature. This work is part of a project where the goal is to develop an active thermal control unit (TCU) capable of regulating a packaged MEMS device at 70 °C ± 0.2 °C in the temperature range from -46 °C to 90 °C. To accomplish this, a thermoelectric cooler (TEC) is sandwiched between the MEMS package and a heat sink with a thermal interface material (TIM). An in-house tool has been used to calculate the optimum TEC parameters, ending up with 72 legs of 1 mm x 1 mm x 2 mm dimension (l x w x h). Simulations in COMSOL show that the TEC can reach the target temperature within the required time and power consumption. In this work we have built a full demonstrator using a thermal test chip (TTC) to represent a MEMS and tested various scenarios in a climate chamber. The tests confirmed to a large degree the simulations of behaviour at constant current. Two regulation algorithms were tested and optimised for of keeping the temperature as constant as possible during temperature changes. The best result kept the TTC temperature within 0.2 °C during temperature cycling. The power consumption during heating mode was reduced by decreasing the heat sink thermal transfer, but such a configuration would not work in cooling mode.**

Keywords— *Thermal managmeent, MEMS, thermoelectric cooler, Peltier effect, Seebeck effect, Joule heating, thermal simulations.*

I. INTRODUCTION

Today, the inertial sensor needs for altitude heading and navigation systems are covered mainly by fiber optic gyrometers (FOGs), ring laser gyrometers (RLGs), Hemispheric Resonant Gyros (HRGs) and mechanical accelerometers. These sensors respond to the metrological needs of less than 1 mg (g - gravitational unit) for accelerometers and less than 1°/h for gyros. Nevertheless, demands for smaller size, weight, power, all at a lower price continue to motivate development of next generation inertial sensing technologies. Continuous improvement of silicon-based MEMS inertial sensor has been achieved over the past years [1,2]. They are now emerging as a new alternative for high accuracy applications.

Whether the detection principle is based on the measurement of a displacement or on the change in frequency of a vibrating element, MEMS-based inertial sensors are sensitive to temperature, due to evolution of stress and material properties in temperature. A large part of the effect of temperature on sensor output can be compensated by modeling of the output versus input temperature and inertial stimulus. After modeling, an error term remains, which cannot be compensated by modeling versus temperature. For the most demanding inertial sensor performance, thermoregulation of the MEMS transducer would be a mean to reduce the magnitude of this error term.

In the PACKOOL project, an active Thermal Control Unit (TCU) is being developed to provide a stable temperature for MEMS devices, T_{MEMS}. The target for T_{MEMS} was set at 70 °C ± 0.2 °C. The main challenge is the large temperature range under which the sensor is expected to deliver the full metrological performance, typically -40 °C to +90 °C. In addition, the SWaP (Size Weight and Power) added by the TCU must remain limited.

A natural candidate for this purpose is a Thermo-Electric Cooler (TEC) which converts electrical flow into thermal flow by utilizing the Peltier effect. They are often used to cool small surfaces or devices [3-5]. To dissipate the heat generated by the TEC they are connected to a heat sink.

TEC devices made from bismuth-telluride as active materials have interesting characteristics to achieve performance and SWaP goals set forth. They are small size, are a mature and widely available technology (commercialized by several companies) and they have the capability to provide both heating and cooling with the same device. This paper addresses the TEC sizing and its assembly within a demonstrator designed to evaluate the thermoregulation performance at MEMS level.

First we present the technical build of the demonstrator. Thereafter we have calculated the optimum architecture of the TEC and simulated the performance to see if it has the potential to meet the specifications set in the project. Finally, we present the assembly of a demonstrator and some thermal testing and results.

II. EXPERIMENTAL

A. Demonstrator Simulation Model

The top left of Fig. 1 shows the conceptual illustration of the inertial sensor targeted for temperature control in this work. The MEMS to be temperature controlled is packaged in a vacuum sealed LCC32 ceramic package with a kovar lid attached using AuSn solid-liquid interface diffusion (SLID) bonding. The lid is attached with a thermal interface material (TIM) to the TEC. The

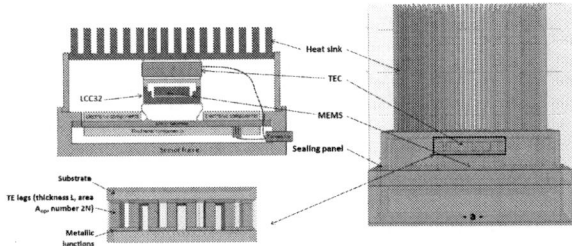

Fig. 1. Top left: conceptual illustration of the inertial sensor to be thermally controlled in this work. Right: The model used in simulating TEC performance. Bottom left: close-up of the model of the TEC used in modelling.

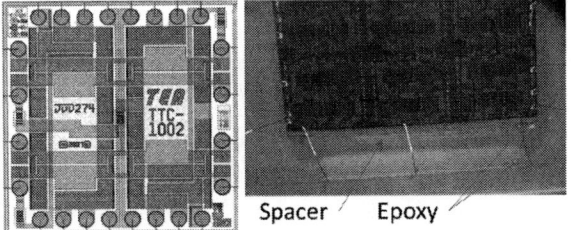

Fig 2. Left: The general layout of the TTC unit cell with pad locations. Right: the 3x2 TTC array mounted on a spacer in the ceramic package.

TEC is in turn attached with a TIM to the heat sink. This is the main thermal path for temperature control. The thermal transport of the wires to the circuit substrate is considered to be negligible.

To the right in Fig. 1 one can see the simulation model based on detailed dimensions from Thales. The behaviour of the TEC is simulated from calculating the Seebeck effect that occurs when a known current density is applied. Therefore, different TEC architecture (e.g. number of TE legs) will give different behaviour.

B. TEC modelling

CEA has developed a dedicated home-made sizing tool to calculate the potential TEC performance based on inputs of number of junctions, leg size, leg thickness, substrate size etc. while considering all specifications already defined by the use case (total volume available, integration constraints, etc.). The resultant geometry of the TEC is then fed into the model (Fig. 1) of the TEC. The overall performance is thus simulated using finite elements modelling (FEM) with COMSOL Multiphysics software. The main specifications considered in the TEC modelling are listed in Table I.

TABLE I.   SPECIFICATION USED TO MODEL THE CORRECT TEC PARAMETERS

Specification	Value
MEMS operating temperature (T_{MEMS})	70 °C
Minimum ambient temperature (T_{amb_min})	-46 °C
Maximum ambient temperature (T_{amb_max})	90 °C
Maximum time to reach operating temperature	60 s

Usually, the geometry of a TEC device is optimized to deliver the best performances for a relatively constant T_{amb}. However, in this case the objective is that the TEC should keep T_{MEMS} at 70 °C at both T_{amb_min} and T_{amb_max}. This means that the TEC geometry cannot not be optimized for either one of these two ambient temperatures but will be adapted to operate successfully in both.
The TEC modelling returned the values given in Table II. The performance based on these parameters was simulated and is discussed in the results section.

TABLE II.   MAIN TEC GEOMETRIC PARAMETERS

Parameters	Values
Legs area - A_{np}	1 x 1 mm²
Legs thickness - L	2 mm
Legs spacing	0.6 mm
Legs number - 2N	72
Metallic junctions size	2,6 x 1 mm²
Substrate size	15 x 12 mm²
Maximum current	2.3 A

C. Demonstrator build

1) TTC in package

The demonstrator in this work was made using a TTC to represent the MEMS. The TTC (from Thermal Engineering Associates) is a silicon chip with integrated heaters and temperature diodes, as seen on the left in Fig. 2. For these tests we used a 2x3 array of the TTC-1002 unit cell, which is 2.54 mm x 2.54 mm. The full array therefore is 5.08 mm x 7.65 mm, which was the largest size that could fit into the designated LCC32 pin package.

The thermal resistance from the bottom of the package cavity to the MEMS active element has been given be approximately 2220 K/W. Hence, it was important to have as high thermal resistance as possible between the TTC and the package to

Fig 3. The assemply of the TEC, LCC32 package w/TTC and the Pt-elements. StayStik is used as a TIM between the LCC32, the TEC and the heat sink.

adequately represent the MEMS. Such a high thermal resistance was not possible to obtain by using any standard die attach material. Therefore, a 1 mm high spacer was designed and 3D printed using high temperature resistance epoxy with a Formlabs2 printer. The spacer was attached using EPO-TEK 353ND, a commonly used underfill material. The result can be seen on the right in Fig. 2. The estimated thermal resistance was calculated to be about 1100-1800 K/W, depending on the thermal resistance of the epoxy material. Prior to the vacuum lid seal, the resistance vs temperature behaviour of the TTC was characterised from -60 °C to 120 °C for calibration purposes.

2) Demonstrator assembly

To avoid the problems encountered in previous work [6] the assembly process was changed in order to certify a proper thermal path from the TEC to the heat sink. Unless otherwise stated, the assembly items were attached using StayStik 581, a silver filled electrically conductive film from Alpha Advanced Materials

First, the heat sink was polished to create a flat surface and then attached to the heat sink base in a convection oven at 220 °C for 180 minutes with 57 kPa applied. In parallel, the TEC was attached to the LCC32 package at 190 °C for 70 minutes on a hot plate with 7 kPa applied. Subsequently the TEC was attached to the heat sink base in a convection oven at 190 °C for 120 minutes with 98 kPa applied. After this Pt-elements were attached on the heat sink base and on the space left beside the LCC32 package to measure T_{HS} and T_{TEC}, respectively, as seen in Fig 3. These were attached using H77 thermal epoxy from EPO-TEK, cured at 150 °C for 60 minutes. Wires were then soldered to the LCC32 package and the Pt-elements. The wires were then stripped and soldered to a custom designed PCB that was again soldered to wires for external connection. to an Agilent 34972A datalogger for temperature measurements and an ITECH IT6000 Bipolar power supply for TEC control.

D. Thermal tests

Thermal cycling was done in an Angelantoni Sunrise 250C climate chamber. The humidity was held at 10%RH at all relevant temperatures.

The bipolar power supply was controlled by a software written in LabWindows to determine the desired applied voltage. We tested two types of regulation algorithms. The first was the standard continuous time PID (C-PID). The second was a discrete time PID regulation using Backward Euler methodology (D-PID), as described in [7] and will not be described here. The latter only uses the three last measurement points and the two output values to calculate the next output. In all tests the applied voltage was limited to 5 V to avoid damage to the assembly.

III. RESULTS

A. TEC performance at constant current

Initial simulations were done in order to certify that the TEC was able to realize a T_{MEMS} of 70 °C in the two values of T_{AMB} in Table 1 by applying a constant current (I_{TEC}) to the TEC and setting the transfer coefficient to air at 5 W/m² K. These were also compared to experimental tests done with the demonstrators to certify that the simulations were representative of the system as a whole.

In the heating mode (T_{AMB} = -46 °C) I_{TEC} was set to 1, 1.5 and 2 A, T_{MEMS} reached 70 °C was in 33 s and 66 s for I_{TEC} of 2 A and 1.5 A, respectively, as seen in Fig. 4, clearly showing that the set temperature could be reached in these conditions. This was also verified experimentally by letting the demonstrator reach -46 °C and then applying I_{TEC} between 1 A and 1.5 A. We did not apply 2 A as it would pose a risk of damaging the demonstrator. As seen in Fig. 4, I_{TEC} of about 1.07 A stabilized close to 70 °C.

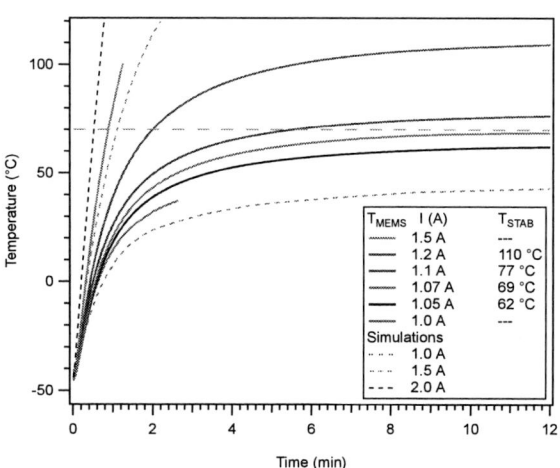

Fig. 4. Evolution of T_{MEMS} at T_{AMB} = -46 °C for different currents applied to the TEC. An I_{TEC} of 1.07 A gives a stablilised T_{MEMS} close to the target of 70 °C. The simulation was done at I_{TEC} = 1, 1.5 and 2 A, and showed that a T_{MEMS} of 70 °C could be reached within the current limitation of the TEC.

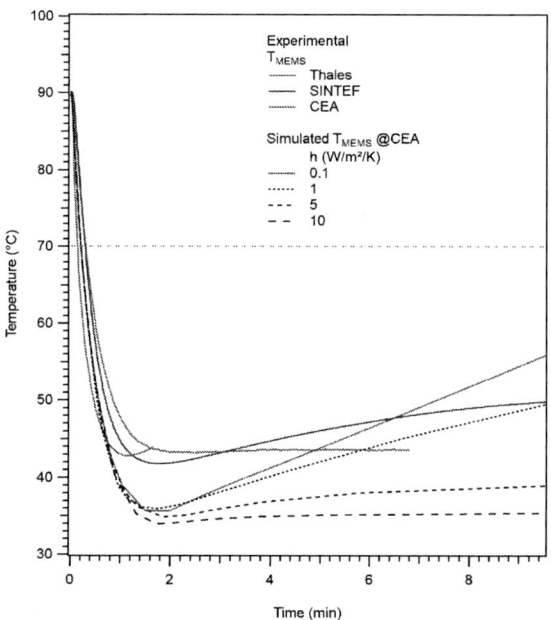

Fig. 5. Simulated and experimental time dependence of T_{MEMS} with T_{AMB} = 90 °C current I_{TEC} = 1 A. Simulations were done with various heat transfer coefficients (h).

In Fig. 5 one can see the performance of the demonstrator detailed above and two similar demonstrators made by Thales and CEA, in the cooling mode with $T_{AMB} = 90 \,°C$. Simulations with four values for the heat transfer coefficient (h) is also shown. I_{TEC} was set to 1 A. All temperature profiles follow the same path for about 30 s, suggesting that the temperature during this period is mostly governed by the thermal mass of the heat sink and not the thermal transfer. The three demonstrators are in fair agreement and the discrepancies are probably due to minor differences in the builds. The two demonstrators from SINTEF and Thales stabilized at a temperature difference, ΔT_{TEC}, of about 56 °C while the one from CEA reached about 63 °C, which is close to the theoretical limit of 72 °C given by the supplier. In all cases, $T_{MEMS} = 70°C$ was reached after 13-18 s, hence the requirement to reach this temperature after 60 s is realistic.

The increase in temperature after the minimum is reached is mainly due to a combination of joule heating in the TEC and the heat sink gradually increasing in temperature while the ΔT_{TEC} is constant.

The simulation results suggest that the specified TEC has the potential to hold $T_{MEMS} = 70 \,°C$ in both heating and cooling mode and that the model used is a good representation of the demonstrator.

TABLE III.   TEC SIMULATION RESULTS

Parameter	$T_{AMB} = -46 \,°C$	$T_{AMB} = +90 \,°C$
I_{TEC}	2 A	1 A
$\Delta T_{TEC\ max}$	N/A	63 °C
$T_{cold\ stab}$	40 °C at t ≈ 7500 s	39.2 °C at t ≈ 2000 s
$T_{MEMS\ stab}$	N/A	40 °C at t ≈ 2000 s
$t_{70\ °C}$	33 s	13 s

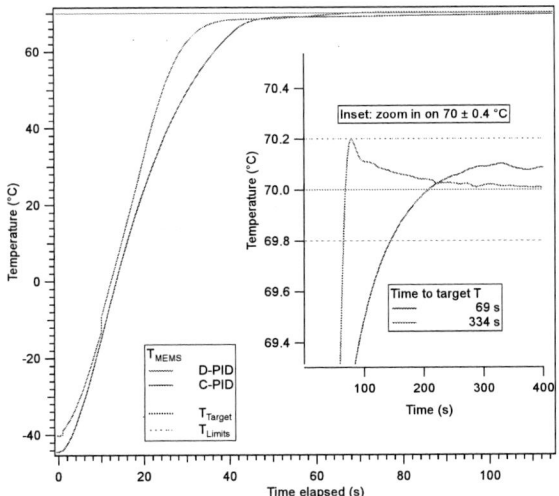

Fig. 6 Response time of the demonstrator using three sets of regulation parameters. The standard C-PID could only reach the target temerature after 143 without overshooting. The best D-PID regulation managed in 65 s.

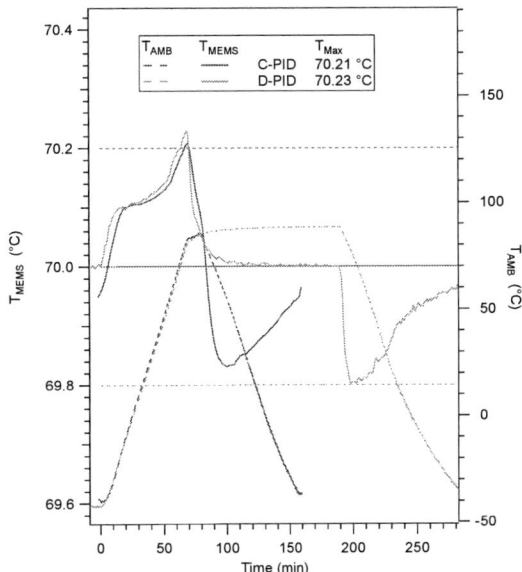

Fig. 7 Temperature stability of the MEMS during thermal cycling between – 46 °C and + 90 °C at ±2 °C/min, for both C-PID and D-PID. There is a slight peak above 70.2 °C

B. Demonstrator performance

In the subsequent sections, the first two tests include the performances of both C-PID and D-PID, while the later tests only use the D-PID regulation parameters that performed best in the two mentioned sections.

1) Response time

In this test the MEMS was stabilized first at $T_{AMB} < -40 \,°C$ to check the time it took to reach 70 °C without an excessive overshoot. Fig. 6 shows the results for the best performing parameters for both C-PID and D-PID algorithms. As a guide to the eye the target is shown as 70 ± 0.2 °C. The D-PID reaches 70 °C after 96 s, outperforming the C-PID by 265 s. Both algorithms show minimal overshoot.

2) Temperature stability during thermal cycling

The aim for the temperature regulation was to keep T_{MEMS} at 70 ± 0.2 °C during thermal cycling from -46 °C to 90 °C at a rate of ± 2 °C/min. In Fig. 7 one can see T_{MEMS} during thermal cycling with both C-PID (blue lines) and D-PID (green lines). Both lines have been smoothed for clarity. The two experiments were done with different dwell times. In essence both regulation algorithms manage to keep T_{MEMS} basically within 70 ± 0.2 °C. In both cases the peak temperature occurs a few minutes after T_{AMB} has passed 70 °C, so the system requires some efforts to reverse the I_{TEC} to go from heating to cooling mode.

3) Power consumption

The power consumption of the TEC during heating mode was approximately 3.5 W, which was above the target for the application, set at 2 W. It was therefore decided to do tests where the thermal transfer from the TEC to the ambient was reduced to minimize the power consumption. In Fig. 8 one can see the five configurations that were tested, with (a) being the benchmark demonstrator. In (b) we added some black ESD foam in between

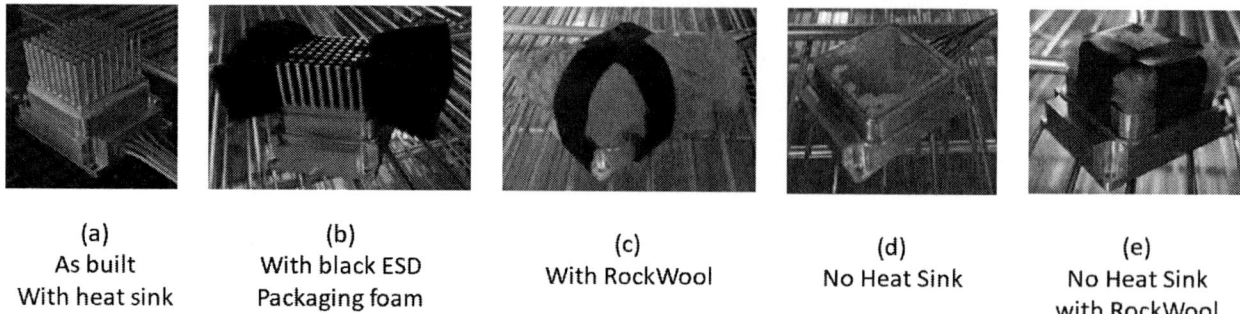

(a)	(b)	(c)	(d)	(e)
As built	With black ESD	With RockWool	No Heat Sink	No Heat Sink
With heat sink	Packaging foam			with RockWool

Fig. 8 Evlolution of temperatures and electrical parameters for a cooling mode test where the initial temperature is 90 °C. TMEMS stabilises at 70 °C after 5 minutes and is held there for about 35 minutes, after which it rises due to insufficient heat dissipation.

the heat sink pillars. For (c) we covered the heat sink in Rockwool insulation material. For (d) we removed the heat sink and in (e) the heat sink base was covered by 1 cm Rockwool. In Fig 9 one can see the power consumption of these configurations during thermal cycling. There is a steady decrease in power consumption from (a) to (e), but in reality only configurations (a) and (d) are realistic, the last of which decreased the consumption from 3.5 W to 2.8 W. The performance during thermal cycling for (a) and (d) was shown to be identical.

Although configuration (d) managed to reduce the power consumption in heating mode, simulations suggested that removing the heat sink would cause the MEMS to overheat in the cooling mode. The reason for this not happening in Fig. 9 is that the experiment was done with forced convection due to the fan in the climatic chamber. To verify this, a simulation was done without the heat sink attached and the current applied to

the TEC in configuration (d) during cooling mode (0.44 A). The thermal transfer coefficient was then varied until the T_{MEMS} was kept at 70 °C. The simulation required a heat transfer coefficient of 11 W/m²/K, which signifies forced convection [8]. Therefore, even though the configuration without heat sink reduced the power consumption in the heating mode, it might not have the needed thermal transfer to operate in cooling mode.

IV. CONCLUSIONS AND FURTHER WORK

In this work we have proposed using a Thermoelectric cooler for thermal management of a MEMS device and investigated its performance. The TEC was sandwiched between the MEMS package and a heat sink and incased in an aluminum housing.

Simulations indicated that the TEC should manage to reach a T_{MEMS} at 70 °C for T_{AMB} ranging from -46 °C to +90 °C. Comparisons with experiments showed that the simulation model was representative of the physical demonstrator.

Two regulation algorithms were tested and compared, a standard continuous time PID (C-PID) and a discrete time PID (D-PID) implementing Backward Euler methodology. The D-PID showed the best response time, 96 s vs. 334 s, while both algorithms showed great performance in temperature stability during thermal cycling.

Efforts to reduce the power consumption during heating mode managed to lower the it to 2.8 W by removing the heat sink. However, in such a configuration the demonstrator would need forced convection to operated successfully in cooling mode.

Further work will include certifying that temperature control of an actual MEMS sensor results in improved output stability.

ACKNOWLEDGMENT

The authors would like to thank the European Defence Agency (EDA) for the support to this work in the context of the project entitled "PACKAGING AND COOLING ADVANCED TECHNOLOGIES (PACKOOL)" funded by France and Norway and coordinated by Thales AVS in the frame of the Project no B-1504-IAP1-GP of the European Defence Agency.

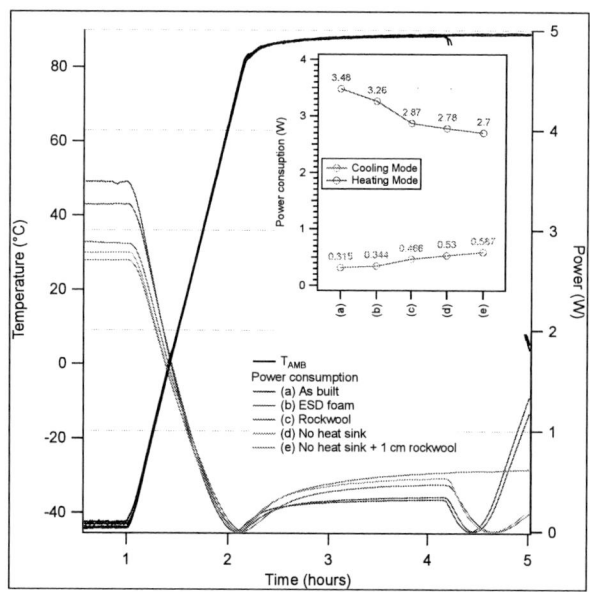

Fig. 9 Power consumption during thermal cycling for the difference configurations in Fig. 8. The lowest power consumption in heating mode reached was 2.7 W, however, just removing the heat sink gave 2.8 W.

REFERENCES

[1] N. Vercier, B. Chaumet, B. Leverrier and S. Bouyat, "A new Silicon axisymmetric Gyroscope for Aerospace Applications," 2020 DGON Inertial Sensors and Systems (ISS), 2020, pp. 1-18, doi: 10.1109/ISS50053.2020.9244886

[2] O. Lefort, I. Thomas and S. Jaud, "To the production of a robust and highly accurate MEMS vibrating accelerometer," 2017 DGON Inertial Sensors and Systems (ISS), 2017, pp. 1-19, doi: 10.1109/InertialSensors.2017.8171494.

[3] K. Fukutani, 'Solid-State Microrefrigerator on a Chip', *Electronics Cooling*, Aug. 09, 2006. https://www.electronics-cooling.com/2006/08/solid-state-microrefrigerator-on-a-chip/

[4] Y. Ezzahri, G. Zeng, K. Fukutani, Z. Bian, and A. Shakouri, 'A comparison of thin film microrefrigerators based on Si/SiGe superlattice and bulk SiGe', *Microelectron. J.*, vol. 39, no. 7, pp. 981–991, Jul. 2008, doi: 10.1016/j.mejo.2007.06.007.

[5] Y. Ezzahri, G. Zeng, K. Fukutani, Z. Bian, and A. Shakouri, 'A comparison of thin film microrefrigerators based on Si/SiGe superlattice and bulk SiGe', *Microelectron. J.*, vol. 39, no. 7, pp. 981–991, Jul. 2008, doi: 10.1016/j.mejo.2007.06.007.

[6] D. N. Wright *et al.*, "Thermal management of MEMS element with thermoelectric-cooler," *2021 23rd European Microelectronics and Packaging Conference & Exhibition (EMPC)*, 2021, pp. 1-5, doi: 10.23919/EMPC53418.2021.9584996.

[7] 'P. Podržaj, "Contionuous VS discrete PID controller," *2018 IEEE 9th International Conference on Mechanical and Intelligent Manufacturing Technologies (ICMIMT)*, 2018, pp. 177-181, doi: 10.1109/ICMIMT.2018.8340444.

[8] https://www.engineersedge.com/heat_transfer/convective_heat_transfer_coefficients__13378.htm

High Aspect Ratio Through-Glass Vias as Heat Conductive Element

Kevin Kröhnert*, Markus Wöhrmann*, Michael Schiffer*, Christian Kelb[†], Norbert Ambrosius[†], Parnika Gupta[‡],
Padraic E. Morrissey[‡], Peter O'Brien[‡,] Martin Schneider-Ramelow[§]

*Wafer Level System Integration
Fraunhofer IZM, Berlin, Germany
Email: kevin.kroehnert@izm.fraunhofer.de
[†]LPKF Laser & Electronics AG, Garbsen, Germany
Email: Norbert.ambrosius@lpkf.com
[‡]Tyndall, Cork, Ireland
Email: parnika.gupta@tyndall.ie
[§]Technical University Berlin, Berlin, Germany

Abstract— In this work metallized Through Glass Vias (TGVs), manufactured by Laser-Induced Deep Etching (LIDE), will be presented as a means to modify the thermal conductivity of glass wafers. Glass, compared to silicon, exhibits superior electrical properties like for e.g. dielectric loss. It has been shown to be a suitable material for hermetic packaging of devices such as microwave emitters. However, the low thermal conductivity of a glass package proves a challenge for high power and continuous operation of active devices due to temperature buildup. The heat which is generated can not be dissipated through the glass as good as silicon (Thermal conductivity of: Silicon 150W/mK; Glass 1,2W/mK). To increase the thermal conductivity of glass packages and to solve issues regarding the operational temperature of ASICs inside a glass package, arrays of copper metallized TGVs can be used as thermal head spreader. Copper has a superior thermal conductivity of about 400W/mK. Aside of using TGVs for signal transmission, high dense TGV arrays are utilized for local increased thermal heat transfer. The benefit of these local heat transfer solution is the avoidance of any thermal cross talk in case of a multi-chip package which is beneficial for temperature sensitive devices like photodiodes, VCSEL or DRAM. Here, we demonstrate how an array of solid-copper-filled TGVs improves the thermal conductivity of a glass wafer both in simulation and experiment. The manufacturing process of the TGVs is illustrated swiftly with reference to prior publications on the subject. The method of copper-filling using a combination of physical vapor deposition and electroplating is explained in more detail. We present the results of our heat transfer simulation and compare them to the obtained experimental results.

Keywords— *Through Glass Via; TGV; LIDE; Interposer; thermal, simulations)*

I. INTRODUCTION

The mayor applications of glass interposers and glass packaging is mayorly driven by high frequency or photonic packages [1][2][4][5][6]. Glass offers superior properties and characteristics to make it feasible for photonic packages and as a base platform to directly mount ASICs, PICs or other components. Over the past couple years, the possibilities regarding the TGV/Cavity fabrication on glass wafers has developed to a point that the limits regarding the geometry of the wafer and the TGVs is determined by the ability to fill TGVs with increasing aspect ratio and wafer thicknesses. Due to the matured technology for glass interposer, which enables the fabrication of complex packages, new challenges emerged. One of the big challenges is the thermal management of active components on glass interposers. Compared to silicon, glass has a ~100 times lower thermal conductivity. To solve this issue copper filled thermal TGVs might be the solution. In this paper the technological possibilities regarding glass interposers with TGV are introduced and the performance of thermal TGVs as heat conductive elements will be presented.

A. Via Fabrication via LIDE

Through-glass-vias (TGV) are a key component of many glass-wafer applications, foremost as interposers in 2.5d chip stacking applications [7]. Currently, a multitude of methods to achieve such vias are under investigation [8], such as microdrilling, powder blasting, laser ablation, deep reactive ion etching or combinations thereof.

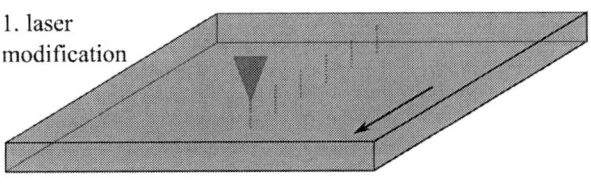

1. laser modification

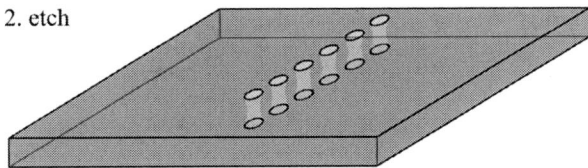

2. etch

Figure 1: Processing steps of creating TGVs in glass wafers. Step 1: Modification, Step 2: Etching of the glass

Laser induced deep etching (LIDE) used here combines a laser modification step (see fig. 1, top) with subsequent chemical etching (fig. 1, bottom) to achieve TGVs with aspect rations of up to 10, depending on the glass type. Since the method is explained comprehensively in previous publications [9] the description given here mainly centers on achievable modifications. Depending on focal parameters, glass type and etch procedure, the final TGV shape can be trimmed in depth, size, and form, allowing to include different functional structures on a wafer with just the two described steps. Figure 2 shows typical cross-sections after etching. The opening angle α varies with the processed material, for BF33 it is ~1.5°. By changing the focal position of the laser either TGVs (fig. 2, a) or blind vias - structures with a rounded or sharp bottom at a defined, tunable depth (fig.2, c) can be defined. Finally, both modification types can be chained or arranged in an array to receive either cutouts or cavities in the substrate material (fig 2 b and c, respectively).

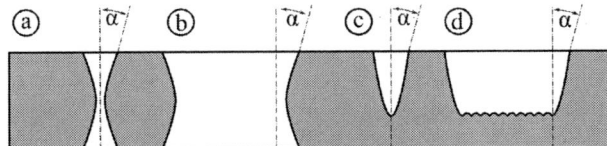

Figure 2. Different types of LIDE modifications. a) TGV, b) chained TGV to create cuttings, c) Blind Vias, d) chained blind vias to create cavities.

For the scope of this paper, TGV as depicted in Figure 2 a) were manufactured using a LPKF Vitrion S5000 fully automated LIDE production system.

B. Interposer Generation and Via Filling

After the TGVs are created the next step is the metallization of such vias. Depending on the via shape and interposer thickness two process flows for the glass interposer fabrication are possible. In Figure 3 an overview of those flows can be seen. For glass interposers with thicknesses below 300µm a carrier wafer is necessary to handle the wafer during processing (Figure 3: A). For thicker glass interposer the processing can be carried out without using a carrier (Figure 3: B). For interposer thicknesses below 300 µm it is ideal to start with a thick glass wafer which has blind vias already etched into the glass. The via filling, the formation of the redistribution layer (RDL) and the fabrication of pads on one side can be performed without using a carrier wafer. After the RDLs and one pad layer is fabricated a carrier wafer is bonded to the frontside and the glass is thinned down to the desired thickness. Now the other side of the wafer can be processed with the desired RDLs and pad/bump metallization. After the processing is completed the carrier wafer will be de-bonded, cleaned and diced. For thicker interposer no carrier is necessary. Therefor no bonding and grinding is necessary. Depending on the application and the requirements the shape of the Via can be different (Tapered – V-Shaped or Hour-glass shaped).

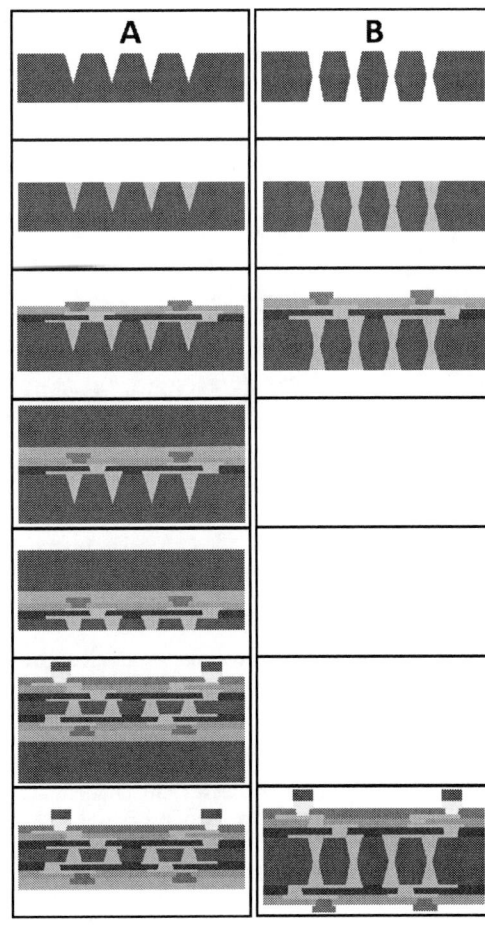

Figure 3: Different processing flows, depending on the wafer thickness (A: glass thickness<300µm; B: glass thickness>300µm).

Both process flows have in common that the challenge is typically the filling or liner plating of the TGVs after the cavities and vias are created with LIDE. Depending on the via geometry and wafer thickness this is done by depositioning an adhesion layer based on titanium and a copper seed layer. For V-Shaped TGVs, like the blind vias in flow a one-sided seed layer deposition is sufficient. The capability of the seed layer deposition is the main limiting factor for the aspect ratio. For hour-glass shaped vias the deposition has to be done for both sides (B). Due to a typical higher aspect ratio and the waist in the center of the TGV. After the seed layer is applied, the following electroplating process uses the thin sputtered layer to bottom up deposit a copper layer in the desired thickness in and on the wafer. It is possible to just form a liner, to fully fill the via and to hermetically fill the holes. Depending on the thickness and the diameter of the TGVs, different options are possible to realize. The liner plating is typically the best option, due to highly reduced plating times and lower induced stresses due to the high CTE of copper. To still have a hermetically sealed TGV, it can be filled with a polymer like substance or closed with a copper membrane. The fully filled via is possible when the diameter is small enough and the copper is necessary (Thermal conductivity). Typical wafer thicknesses range from ~50-500µm and via diameters from 10-100µm. In Figure 4 some examples of glass interposers with different thicknesses, different metallization concepts, via geometries and diameters can be seen.

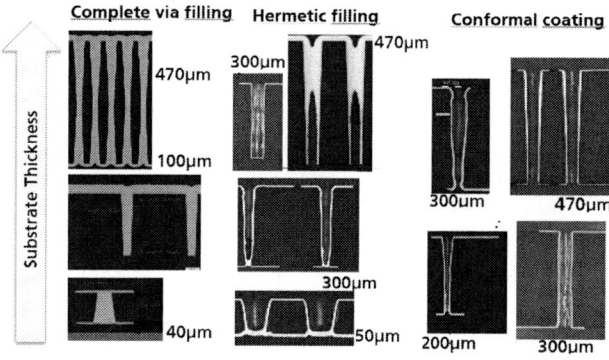

Figure 4: Different realized TGVs with various shapes, sizes and metallization concepts by Fraunhofer IZM. Substrate thickness is mentioned on each TGV.

After the electroplating is done, chemical-mechanical polishing removes the excess copper from the wafer surface, only leaving the copper inside the TGVs. After the Vias are finished the RDL and Pad metallization can be fabricated. Due to the possibility of having cavities and holes) on the wafer surface the following RDL processes are challenging due to some limitations regarding the choice of processes and materials which can be used. An overview of the possibilities, regarding layer count, passivation and mentalizations can be seen in the following table.

Table 1. Glass Interposer Specifications

Glass interposer specifications	
Layer Count Top/Bottom	3/3 + Pads
Line/Space µm	8/8
Aspect Ration	10/1
TGV Pitch	1.2xTGV Diameter
Interposer Thickness	40-500µm
TGV Diameter	10-1000µm

C. Packaging and Testing

To verify the performance of the copper filled TGV a test-setup using a laser diode bonded to a glass interposer with thermal TGVs is created. A FLIR thermal microscope will be used for the measurement of the maximum surface temperature of the laser die. In Figure 5 an overview of the test setup can be seen. The topside contact of the laser die which was on the opposite side of the wire-bonded contact had black paint to reduce measurement uncertainties by increasing surface emissivity.

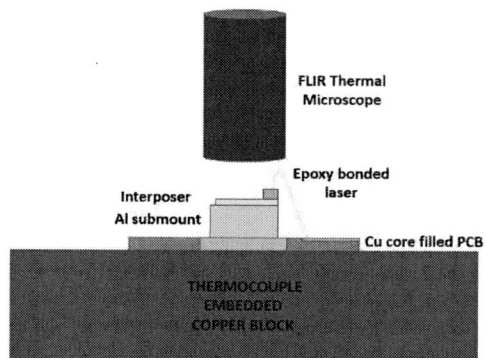

Figure 5. Measurement Setup for IR thermal Imaging of a glass interposer with a bonded laser diode

D. Thermal Simulations and Assumptions

To compare the measured thermal results to the simulated results COMSOL was used to create an accurate model of the TGV array and the glass interposer. An accurate model of a single unit TGV cell was designed. Based on the copper in the simulation these unite cell shows different thermal conductivity in x/y and z direction. The effective thermal conductivities were simulated for a unit TGV cell, where k_{xy}= 2.27 W/m.K and k_z= 94.78 W/m.K. This effective thermal conductivity in z-direction was approximately 78 times higher than that of glass (1.2 $Wm^{-1}K^{-1}$). The effective thermal conductivity was then used to model the laser die bonded on glass with TGVs.

II. EXPERIMENTAL

To investigate the thermal behavior of a glass interposer and to evaluate how the TGVs would perform regarding thermal conductivity a test vehicle was fabricated. The test vehicles have an array of TGVs, which are then filled with copper to transfer the heat through the glass away from a laser, which is used as a heat source. The target application of such a glass interposer is an optical package with cavities and thermal/electrical TGVs. To create cavities with a thickness of around 250µm and still have enough mechanical stability the glass substrate needs to be thicker than 400 µm and TGVs with a high aspect ratio in the range of 1:10. The via diameter should be as small as possible so metallization is still possible (Aspect Ratio limits the size). The fabrication of such glass wafers with the LIDE technology is typically not the limiting factor. For this project an interposer thickness of ~460µm with hourglass shaped TGVs was chosen. The distance between each TGV was defined with only 10µm to have a dense array of TGVs and a high copper amount inside the glass.

A. Fabricated Glass Wafers with TGVs

Several wafers were produced using LIDE, with TGVs in different diameters and spacings. The raw material used were 8" double-side polished wafers (Schott BF33) with a thickness of 500 µm.

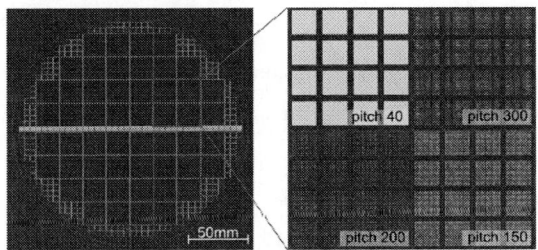

Figure 6: Overview and Detail of one of the produced wafers.

All produced wafers were designed with square TGV patterns with different pitches between 40 µm and 300 µm, as shown in Figure 6:. After laser modification, the wafers were etched to 40 µm, 50 µm, and 60 µm TGV diameter. For the different TGV diameters, the minimum pitch was adjusted accordingly to allow for a minimum clearance of 10 µm between the individual TGVs. Table 2 shows an overview of the different wafer designs. For the wafer with the smallest pitch pattern (design #1), the total number of TGVs is approximately 3.2 million.

Table 2. Design Overview

Design	Diameter	Pitch Patterns
#1	30 µm	40 µm / 150 µm / 200 µm / 300 µm
#2	40 µm	50 µm / 150 µm / 200 µm / 300 µm
#3	50 µm	60 µm / 150 µm / 200 µm / 300 µm

To verify the TGV diameter and position, a design #1 wafer was measured using a Werth Videocheck coordinate measurement machine. For 1994 randomly selected positions, the mean diameter was 31.7 µm with a standard deviation of 0.46 µm. The positional error was 0.04 µm in X-direction and -0.05 µm in Y-direction with standard deviations of 0.51 µm and 0.41 µm, respectively. Due to the time-consuming measurement process, the other designs were only tested for

compliance with the desired TGV parameters by sampling a small number of TGV using a Keyence VK-X200 Series confocal microscope. Table 3 shows a comparison between the measured parameters for all three wafer designs.

Table 3. Measurement Results for produced TGV wafer designs.

Design	#1	#2	#3
Mean Diameter (µm)	31.7	39.6	48.1
Dia. Std. Dev. (µm)	0.46	0.46	0.45
Mean X-Pos. Error (µm)	0.04	0.59	0.14
X Std. Dev. (µm)	0.51	0.56	0.71
Mean Y-Pos. Error (µm)	-0.05	0.00	0.00
Y Std. Dev. (µm)	0.41	0.34	0.24
Sample Size	1994	59	40

B. Interposer for Thermal Testing

After the glass wafers were fabricated, the wafers were sputtered and electroplated from both sides to achieve fully filled TGVs. The variant with the 40µm diameter TGVs could not be successfully metalized due to the high Aspect Ratio (1:12.5). Therefor we used for further processing the glass interposers with the 50µm vias, which could be successfully filled with copper.

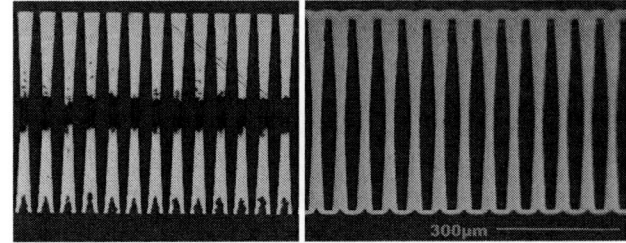

Figure 7: Left: 40µm Via Diameter with unsuccessfully plating; Right: 50µm Via Diameter fully filled with copper.

After the excess copper was removed from, the wafer surfaces the top and bottom of the wafer was metallized with a gold metallization to be able to solder heat sources directly to the interposer for the following testing steps. The wafer was then diced into 4.5x4.5mm² big glass chips.

C. Measured Results

The two packages seen in Figure 8 were built with the laser die bonded on the glass substrate (with and without TGV) using silver epoxy. The 4.5x4.5 mm² glass substrates had a 500 nm gold layer deposited on the top and bottom of the substrates. This subassembly was then mounted on a 5x5 mm² aluminum submount which was then connected to a Copper-core filled PCB. This PCB was placed on a large Copper block (8x8x1 cm³) which was mounted on four 4x4 cm² TECs and the temperature was calibrated at 18°C for reference using the TEC Controller. The TGVs had an hourglass shape with 50 µm top and bottom diameter and a diameter of 25 µm in the middle.

The pitch of the TGVs was 60 μm. The thickness of the glass interposer was 460μm.

Figure 8. Packaged laser die on (a) glass with TGVs (b) glass substrate.

The simulated models for the package with and without TGVs were shown in Figure 9. The maximum surface temperature of the laser die was 207.9°C in case of glass and 44.1°C in case of glass with TGVs at input powers of 186mW and 150.35mW respectively (100mA laser driving current in both cases). The input powers for both structures were calculated from experimentally derived data.

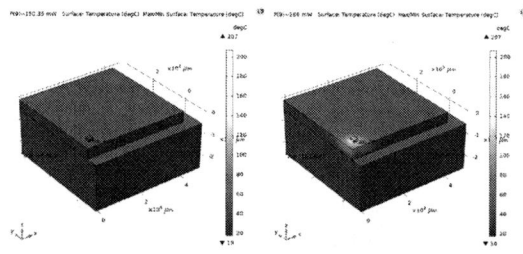

Figure 9. Simulated thermal models of packaged laser die on (a) glass with TGVs and (b) glass substrate.

The thermal measurements were taken using the FLIR Infrared Microscope for the laser dies mounted on both samples (with and without TGVs). The images and maximum surface temperature readings were recorded in Figure 10 and Figure 11.

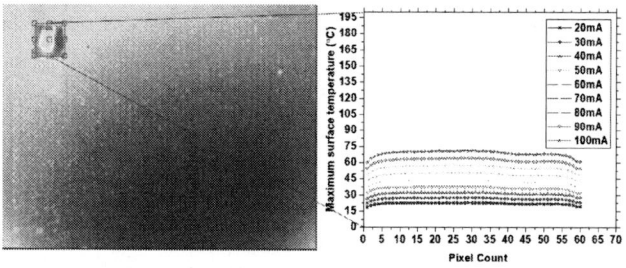

Figure 10. Measured temperature across the vertical section of the laser die on glass with TGVs substrate at different laser driving currents.

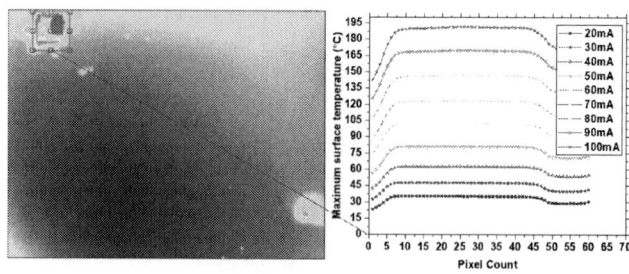

Figure 11. Measured temperature across the vertical section of the laser die on glass substrate at different laser driving currents

The comparison of the measured and simulated in case of glass substrate shows good agreement until the driving current is increased to 80mA, this was observed because of the additional thermal losses at surface interfaces in the package. When we compared the results in case of glass with TGVs, the simulation results predicted a 163°C maximum decrease in temperature with the vias, which was higher than the 120°C decrease in maximum temperature according to the measurement results (compared at 100mA driving current).

This discrepancy between the measured and simulated results was investigated further to reveal the impact of via filling on the heat transfer capability of the TGV substrate. Figure 12 shows the decrease in effective thermal conductivity from 94.78 Wm^{-1}K^{-1} to 10 Wm^{-1}K^{-1} as the via filling decreased from fully-filled to partially-filled. The partially-filled TGV simulation results were in better agreement with the measurement results. The measured and simulated results agreed closely in case of glass. The increasing deviation of measured results from the simulated results as the input power increases can also be partially attributed to the additional thermal losses at component interfaces, thermal conductivity anisotropy and the radiation thermal losses that are not factored in the simulation setup. The via filling factor was approximated in the simulation model using a conical cavity in the via which represents the unfilled via portion as seen in Figure 7. To better understand the impact of via filling on thermal performance of the TGV, the effect of changing the height and bottom radius of the conical cavity on the effective thermal conductivity was detailed in Figure 13, where the increase in height and radius of the conical cavity had a directly proportional impact on the decrease in effective thermal conductivity of the TGV substrate.

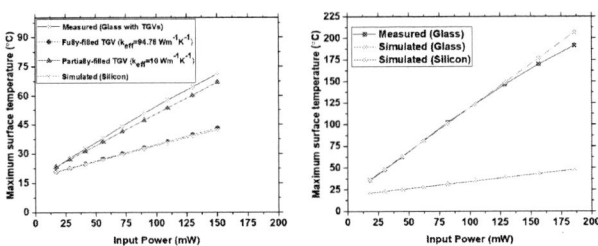

Figure 4. Graphs showing the measured and simulated results in (a) glass with TGVs and (b) glass substrate at different input powers

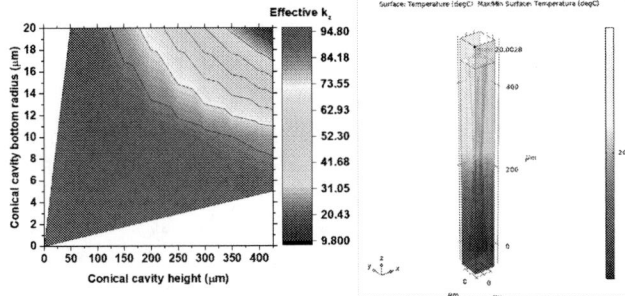

Figure 5. Impact of the changing conical cavity height and radius on the effective thermal conductivity of the glass with TGVs substrate and an optimized model of the via with the conical cavity

III. CONCLUSION

In this work the fabrication of glass interposers and its challenges were presented. One of the disadvantages of glass was the low thermal conductivity. The possibility of using TGV arrays to greatly increase the thermal conductivity of glass was evaluated and measurements on real test vehicles were performed. The acquired data showed that thermal TGVs reduce the surface temperature of assembled devices drastically which makes them suitable as heat conductive elements for glass interposer. The simulation data shows that the thermal conductivity should be even higher compared to the acquired data. To achieve such high values the performance of the via filling processes of such high aspect ratio TGVs in a thick interposer has to be improved to prevent voids inside the TGVs

which will lead to a lower thermal conductivity compared to the simulated data. To further improve the concept of such vias, we will improve the thermal conductivity even further and evaluate the reliability of such dense arrays with copper inside during thermal cycling in the furture.

ACKNOWLEDGMENT

The authors would like to thank all the participating colleagues from Fraunhofer IZM, LPKF Laser & Electronics AG, Tyndall. The work was funded by the European Union (Horizon 2020).

REFERENCES

[1] Kevin Kröhnert, M. Wöhrmann, N. Jürgensen, K. D. Lang, T. Galler, T. Chaloun, C. Waldschmidt, M. Schulz-Ruhtenberg, N. Ambrosius, and R. Ostholt (2019) Through Glass Vias for hermetically sealed High Frequency Application. Additional Conferences (Device Packaging, HiTEC, HiTEN, & CICMT): January 2019, Vol. 2019, No. DPC, pp. 000125-000140.

[2] T. Galler, T. Chaloun, K. Kröhnert, M. Schulz-Ruhtenberg and C. Waldschmidt, "Hermetically Sealed Glass Package for Highly Integrated MMICs," 2019 49th European Microwave Conference (EuMC), Paris, France, 2019, pp. 292-295.

[3] Roman Ostholt, Norbert Ambrosius, "High-throughput via formation in solid-core glass for ic substrates," 2017 MiNaPAD Conference, May 17-18, Grenoble, France.

[4] T. Galler, et al., "Glass Package for Radar MMICs Above 150 GHz," in IEEE Journal of Microwaves, vol. 2, no. 1, pp. 97-107, Jan. 2022, doi: 10.1109/JMW.2021.3122067.

[5] K. Kröhnert, G. Friedrich, D. Starukhin, M. Wöhrmann, M. Schiffer and M. Schneider-Ramelow, "Reliabillity of Through Glass Vias and hermetically sealing for a versatile sensor platttform," 2020 IEEE 8th Electronics System-Integration Technology Conference (ESTC), 2020, pp. 1-6, doi: 10.1109/ESTC48849.2020.9229834.

[6] K. Kröhnert *et al.*, "Versatile Hermetically Sealed Sensor Platform for High Frequency Applications," *2021 23rd European Microelectronics and Packaging Conference & Exhibition (EMPC)*, 2021, pp. 1-8, doi: 10.23919/EMPC53418.2021.9584974.).

[7] Okoro, C., Jayaraman, S., & Pollard, S. (2021). Understanding and eliminating thermo-mechanically induced radial cracks in fully metallized through-glass via (TGV) substrates. Microelectronics Reliability, 120, 114092. https://doi.org/10.1016/j.microrel.2021.114092

[8] Hof LA, Abou Ziki J. Micro-Hole Drilling on Glass Substrates—A Review. Micromachines. 2017; 8(2):53. https://doi.org/10.3390/mi8020053

[9] R. Santos, J. -P. Delrue, N. Ambrosius, R. Ostholt and S. Schmidt, "Processing Glass Substrate for Advanced Packaging using Laser Induced Deep Etching," 2020 IEEE 70th Electronic Components and Technology Conference (ECTC), 2020, pp. 1922-1927, doi: 10.1109/ECTC32862.2020.00300

Reliability Characterization of Graphene Enhanced Thermal Interface Material for Electronics Cooling Applications

Markus Enmark*, Murali Murugesan†, Amos Nkansah†, Yifeng Fu*, Torbjörn M.J. Nilsson‡ and Johan Liu*

*Electronics Materials and Systems Laboratory, Department of Microtechnology and Nanoscience (MC2),
Chalmers University of Technology, Kemivägen 9, SE-412 96 Göteborg, Sweden.
†SHT Smart High-Tech AB, Kemivägen 6, SE-412 58, Göteborg, Sweden
‡Saab AB, Solhusgatan 10, SE-412 76, Göteborg, Sweden
E-mail: johan.liu@chalmers.se

Abstract—Graphene-based products are gaining popularity in thermal management applications in high performance electronics systems. The ultra-high thermal conductivity of graphene together with its relatively low density makes it a suitable material for reaching high cooling capability in lightweight applications. An example of products that are starting to enter the market is graphene enhanced thermal interface materials (TIMs). Pristine graphene enhanced TIMs are well characterized and show high thermal conductivity and low thermal interface resistance. Before these TIMs can take the next step from being a niche product to reach high volume sales on the market, it needs to be proven that they have stable performance over time when conditioned and aged according to industry reliability standards.

In this work, a set of customized test rigs was designed, and graphene enhanced TIMs of three different thicknesses were tested. The TIMs were compressed by 30% and then subjected to three different industry standard reliability tests; thermal aging, temperature cycling and damp heat. The thermal resistance was measured sequentially during each test to monitor change over time. The reliability tests are still ongoing and so far the tested graphene enhanced TIMs have stable performance over time with some observable trends for the different tests. At the current test time the maximum degradation in thermal resistance is 13%, measured after 511 cycles in the thermal cycling test. The used test method is deemed promising for reliability comparison and future requirement standardization on thermal pads.

Keywords—thermal interface material; graphene; reliability testing; electronics cooling

I. INTRODUCTION

Thermal interfaces are an inevitable part of every thermal management solution where heat needs to be transported away from a heat generating device. High performance electronics is one of the areas where thermal management solutions need to be improved to keep up with the increasing power density in chips. In any thermal management system, two uneven surfaces in junction will introduce a thermal resistance. Thermal interface materials (TIMs) is the name of a family of different materials that aim to reduce thermal resistance in the interface between two surfaces. TIMs can be composed of many different materials, in different shapes, compositions and state of aggregation depending on the application. Conventional TIMs can be categorized into six different types including thermal grease, thermal pads, phase change materials, gels, conductive adhesives and solders [1]. Although the TIM categories are

different, they aim to do the same thing, namely fill the air gaps in the interface between two surfaces with a highly conforming and thermally conductive material.

Carbon nanomaterials have been the target for extensive research to be used in TIMs. This is mainly due to their favorable intrinsic properties such as high thermal conductivity, low density and high aspect ratio. Commonly reported thermal conductivity values of graphene is in the range of 2 000 - 5 300 W/mK [2]–[5] and carbon nanotubes between 3 000 - 3 500 W/mK [6], [7]. These values are between 5-13 times higher compared to conventional filler materials such as silver and copper.

Graphene and carbon nanotubes are often used as filler materials in thermal grease and composite thermal pads to increase the thermal conductivity [1], [8]. Additionally, researchers have also focused on carbon nanomaterials as standalone material to connect two surfaces together, such as carbon nanotube arrays, buckypapers, carbon based aerogels and vertically aligned graphene films [9].

The thermal conductivity of a graphene enhanced thermal pad is highly dependent on the orientation of the graphene flakes. A monolayer of carbon atoms facilitates electron and phonon transfer in-plane while the thermal conductivity is worse in the cross-plane direction in a multilayer structure. One way to engineer a graphene enhanced TIM is by combining vertically aligned graphene sheets with a polymer binder. This has been proven to be a successful way of exploiting the exceptionally high thermal conductivity of graphene in thermal pads [10]. For instance, it has resulted in lightweight products that have high thermal conductivity, are compressible and have good conformity to other surfaces [11]–[13]. The performance of these products is well investigated with different thermal characterization methods such as laser flash method according to ASTM E1461-13 and steady state measurements according to ASTM D5470-17 [14], [15].

A common knowledge gap when it comes to new innovative products is prediction of expected lifetime and performance over time. Guo et al. states that future research on carbon based TIMs should be more focused on endurance and reliability in order to reach a broader market [9]. Today there are no standardized test methods for predicting the lifetime of TIMs before they are put into an actual application. It is common that standards for other closely related components have been used for evaluating TIM reliability [16]. Therefore, there is a need for a generic reliability test method that TIM manufacturers can use to evaluate reliability without being dependent on an application from a potential buyer.

When testing reliability of a thermal pad it is important to keep it in a compressed state as similar as possible to the application of interest. It should also be possible to maintain the compressed state throughout the reliability test and still be able to monitor the performance of the TIM. Then it is possible to see when, or if, the TIM performance drops below an unacceptable level.

In this work, customized test rigs were designed for the purpose of testing thermal performance of graphene enhanced thermal pads over time without having to remove them from the compressed state. This method gives an opportunity to condition and measure thermal performance of the TIM without the risk of affecting the test specimen by removing it and putting it back in a compressed state.

II. EXPERIMENTAL

Twelve customized test rigs were manufactured according to Fig. 1. The bottom copper block seen in Fig. 1(a) has room for two cartridge heaters and two thermocouples. The top copper block seen in Fig. 1(b) has room for two thermocouples.

Fig. 1: (a) Bottom copper block with aluminum spacers in each corner and a sample of graphene enhanced TIM placed in the middle. (b) Top copper block with screws to assemble the compression rig. (c) Assembled compression rigs before start of the reliability test.

Graphene enhanced TIMs with the product name GT70S were provided by SHT Smart High-Tech AB in three different thicknesses 0.8 mm, 1 mm and 2 mm. The pads were cut into 30x30 mm pieces and put inside the compression rigs. After cutting, the pad thickness was controlled by a micrometer to be able to determine the compression set of the thermal pad after the reliability tests. The TIM was put in the middle

of the bottom copper block seen in Fig. 1(a) and the rigs were assembled with four screws and aluminum spacers in between the bottom and top copper block to obtain the desired compression rate of 30%. Four rigs were run in every reliability test, one rig for each TIM thickness and an empty rig for reference. The empty rig had the same spacer thickness as the 1 mm TIM and was included in the reliability tests for the purpose of eliminating the contribution of the rig itself.

A. Thermal characterization

Before starting the reliability tests, steady state measurements were carried out to determine the total thermal resistance between the bottom and the top copper block.

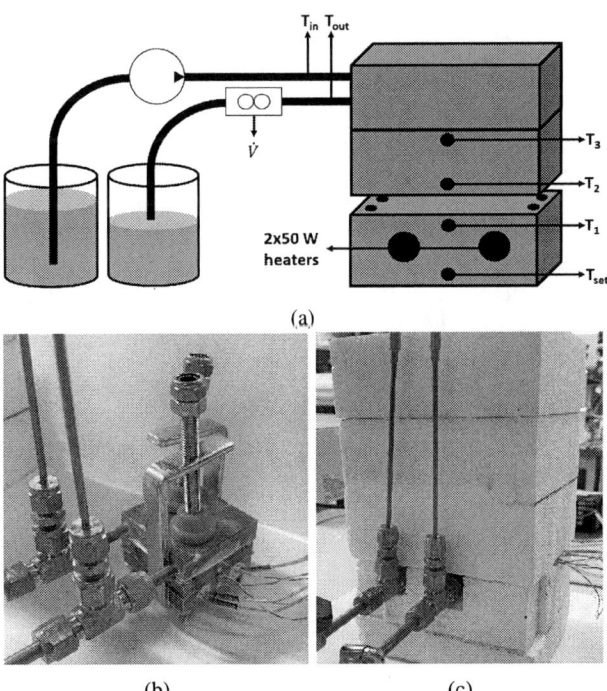

Fig. 2: (a) Schematic picture of the test setup used for thermal characterization. (b) Assembled test setup with attached water-cooled heat sink, heaters and thermocouples. (c) Test setup during operation, insulated with Styrofoam.

The bottom copper block was supplied with a constant power of 100 W by using two 50 W cartridge heaters. A water-cooled heat sink was applied on the top of the rigs by applying thermal grease and clamping it in place with two clamps. The clamp surfaces in contact with the test rigs were covered with a polymer to have a low thermal conductivity between the test rig and the clamps. The tightening torque used was 1.6 Nm for each of the clamps. A schematic of the complete test setup is shown in Fig. 2(a) and an assembled test rig can be seen in Fig. 2(b). When running the thermal characterization test the entire test rig was insulated in Styrofoam to reduce convective cooling by the ambient, this is illustrated in Fig. 2(c).

Each steady state measurement was run for 10 minutes and the temperature in the copper blocks T_1, T_2 and T_3 were continuously sampled. The temperature of the cooling water

T_{in}, T_{out} and the volumetric flow rate were also measured in order to calculate how much of the heat that is carried away by the cooling water. The heat loss to the ambient was negligible and the total thermal resistance could be calculated by:

$$R = \frac{\Delta T}{Q} \qquad (1)$$

where ΔT is $T_1 - T_2$ and Q is the power input in watts.

B. Reliability tests

After thermal characterization of the pristine rigs and thermal pads, three industrial standard reliability tests were started to test different degradation mechanisms and potential failure modes. The reliability tests will be run until a minimum of 1 000 hours, TIM failure or no more observable degradation. At this point in time the reliability tests are still ongoing and data from the thermal characterization is gathered continuously.

1) Thermal aging: The purpose of the thermal aging test is to degrade the polymer binder by increasing the thermo oxidative aging rate and thereby change its mechanical properties and chemical composition. It also accelerates the compression set of the pad and thereby reduces the force exerted on the copper blocks over time.

The thermal aging test was carried out at a constant temperature of 120°C. So far, the test rigs were taken out of the chamber after 258, 580 and 780 hours to do the thermal characterization.

2) Damp heat: The damp heat test introduces moisture as another potentially degrading parameter. Moisture is absorbed by the polymer binder, can cause swelling and change mechanical properties of the composite. It may also chemically degrade both the polymer binder and the graphene-based sheets.

The damp heat test was carried out at a constant temperature of 85°C and a relative humidity of 85%. The test parameters are in accordance with JEDEC standard JESD22-A101D.01 [17]. So far, the test rigs were taken out of the chamber after 228, 480, 790 and 1044 hours to do the thermal characterization tests.

3) Thermal cycling: Thermal cycling is mainly carried out to see if a difference in coefficient of thermal expansion can degrade the TIM over time as the materials that are joined together expand and contracts during the cycling. This has the potential to both introduce shear strain within the thermal pad between the polymer binder and the graphene-based sheets and between the thermal pad and the copper blocks.

The thermal cycling was carried out according to JEDEC standard JESD22-A104F test condition G, between -40°C and +125°C [18]. So far, the test rigs were taken out of the chamber after 239 and 511 cycles to do the thermal characterization tests. The total cycle time was 70 minutes, a shorter cycling time was not possible due to the sample rigs relatively high volumetric heat capacity. Two complete cycles can be seen in Fig. 3.

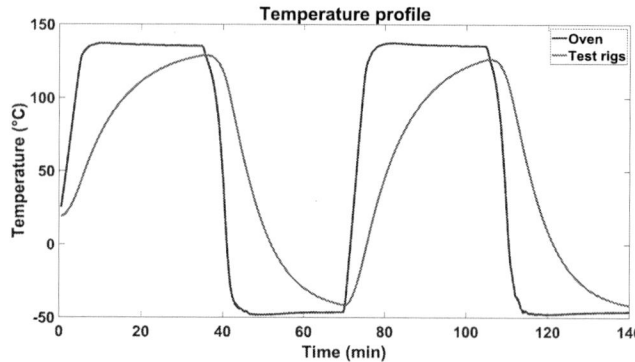

Fig. 3: Temperature profile during the thermal cycling test. Blue line is temperature of the oven and red is temperature profile of the test rigs.

III. RESULTS AND DISCUSSION

When measuring the thermal resistance over the two copper blocks in the test rigs, seen in Fig. 2(a), the result also takes into account the thermal resistance of the aluminum spacers and the screws that connect the two copper blocks. Therefore, it is important to notice that the reported thermal resistances are not the actual thermal resistances of the TIM but a combination of all the components that connects the two copper blocks. In the end of this section the contribution from the reference rigs is subtracted from the total thermal interface resistance to only get the resistance of the 1 mm thick TIMs.

The mean thermal resistances reported in this section are calculated from the steady state measurements between t=200 seconds and t=500 seconds when the thermal resistance has reached steady state. By monitoring how the mean thermal resistance varies as the specimens are conditioned it is possible to observe trends. The mean thermal resistances reported in this section are only based on one steady state measurement per test specimen. To evaluate the repeatability of the steady state measurements three consecutive measurements were carried out on the reference rig and the 1 mm TIM after 900 hours of thermal aging. The measured thermal resistances displayed in Table I have good repeatability for both the reference rig and the 1 mm TIM. In future studies more samples per TIM thickness should be included in every reliability test to get a more statistically assured result.

TABLE I: Test series to test repeatability of steady state measurements. The measurements were carried out after 900 hours of thermal aging.

	Ref rig (K/W)	1 mm TIM (K/W)
Test 1	0.2489	0.0273
Test 2	0.2426	0.0285
Test 3	0.2420	0.0284
Standard deviation	3.8×10^{-3}	6.5×10^{-4}

In Fig. 4 the measured thermal interface resistances are plotted against aging time for the thermal aging specimens. It shows that the thermal interface resistance of the reference rig

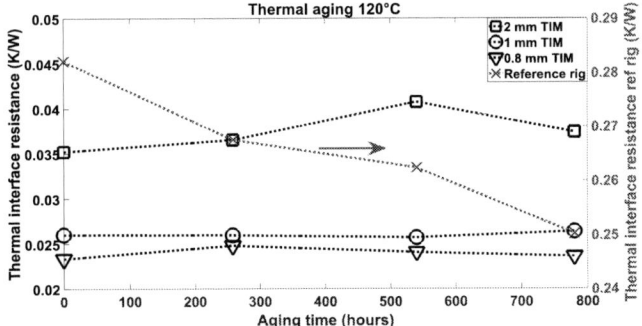

Fig. 4: Total thermal interface resistance plotted against aging time for the thermal aging specimens.

decreased with 11% after 780 hours. The thermal resistances of the rigs containing the 0.8 mm and 1 mm TIM are rather constant over the aging time. The rig with the 2 mm TIM shows a more distinct increase of 16% after 540 hours that decreases to 6% when measured after 780 hours. This is not an expected behavior but can possibly be explained by a higher moisture content of the TIM at 780 hours. The steady state measurements at 780 hours could not be carried out directly after removal from the oven and had to be left in room temperature for several days before being measured. This can result in a higher moisture content in the polymer binder and thereby also a lower thermal resistance. The result highlights the importance of having a standardized time frame for characterization of the test specimen after removal from any conditioning chamber.

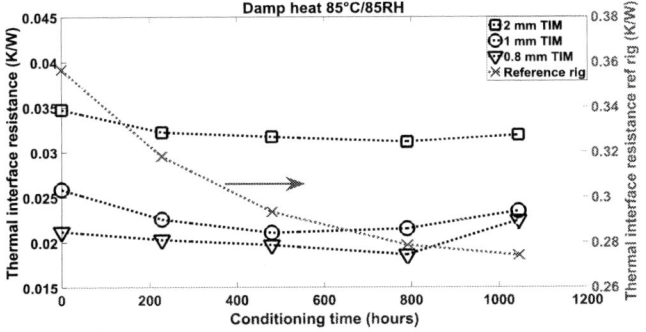

Fig. 5: Total thermal interface resistance plotted against conditioning time for the damp heat specimens.

In the damp heat test, shown in Fig. 5, the thermal interface resistance for the reference rig has decreased with 23% after 1044 hours in 85°C and 85% relative humidity. The thermal resistances also decrease for the 1 mm and 2 mm TIMs with 9% and 8% respectively while the 0.8 mm TIM shows a 6% increase in thermal resistance. An interesting observation is that every test specimen behaves similarly over time. The thermal resistance initially decreases but at some time after 480 hours the trend shifts and the thermal resistance start to increase instead. A possible explanation is that the polymer binder absorbs moisture during the test until it is saturated. An increased moisture content could have a positive effect

on the intrinsic thermal conductivity of the TIM and thereby also contribute to a lower thermal resistance. There is also a possibility that water can fill micro air voids between the TIM and copper surfaces and thereby lower the thermal contact resistance. The trend shifts when the degradation mechanisms of the TIM become greater than the positive effects of moisture, resulting in the shape of the curves seen in Fig. 5.

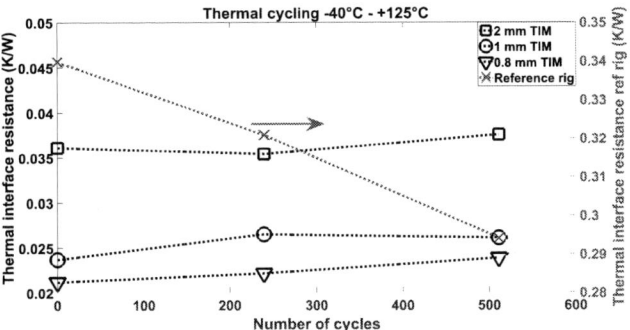

Fig. 6: Total thermal interface resistance plotted against number of cycles for the thermal cycling specimens.

The results from the thermal cycling test are shown in Fig. 6. The trend for the reference rig is the same in the thermal cycling test with a 13% lower thermal resistance after 511 cycles. The 2 mm, 1 mm and 0.8 mm rigs show a change in thermal resistance of +4%, +10% and +13% respectively. There is a slight indication that the thinner TIMs are more susceptible to degradation during thermal cycling. The test is still only in the beginning and more data points for higher numbers of cycles are necessary to confirm this trend.

The reliability tests have been run for different lengths of time and are still ongoing. A compilation of how much the thermal resistances have changed so far in the different reliability tests can be seen in Table II.

TABLE II: Change in thermal resistance from pristine rigs to current conditioning time.

	Thermal aging 780 h	Damp heat 1044 h	Thermal cycling 511 cycles
Ref rig	-11%	-23%	-13%
0,8 mm TIM	+1%	+6%	+13%
1 mm TIM	+1%	-9%	+10%
2 mm TIM	+6%	-8%	+4%

The reference rigs that were included in every reliability test show a general trend of decreasing thermal resistance. The trend is most significant in the damp heat test but can also be observed in both thermal aging and thermal cycling. This means that there is a small negative contribution in thermal resistance on the TIM containing test rigs. Although not completely investigated at this point, it is believed to be due to diffusion bonding between the copper blocks and the aluminum spacers. This will be analyzed when the test rigs are opened at the end of the reliability tests. In hindsight it would have been better to use ceramic spacers on all test rigs to avoid any diffusion bonding. Ceramic spacers would also be beneficial due to lower thermal conductivity and thereby have a lower contribution to the total thermal interface resistance.

Since the reliability tests have been run with a reference rig with the same spacer thickness as the 1 mm TIM, it is also possible to calculate the actual thermal interface resistance of the TIM by subtracting the contribution from the reference rig. The following equations of parallel thermal resistance are used.

$$\frac{1}{R_{\text{tot}}^{\text{Ref}}} = \frac{1}{R_{\text{spacers}}} + \frac{1}{R_{\text{air}}^{\text{Ref}}} \tag{2}$$

$$\frac{1}{R_{\text{tot}}^{\text{TIM}}} = \frac{1}{R_{\text{spacers}}} + \frac{1}{R_{\text{air}}^{\text{TIM}}} + \frac{1}{R_{\text{TIM}}} \tag{3}$$

$$R_{\text{TIM}} = \frac{1}{\frac{1}{R_{\text{tot}}^{\text{TIM}}} - \frac{1}{R_{\text{spacers}}} - \frac{1}{R_{\text{air}}^{\text{TIM}}}} \tag{4}$$

From the reference rig measurements, we get $R_{\text{tot}}^{\text{Ref}}$ and can calculate R_{spacers} and $R_{\text{air}}^{\text{Ref}}$ seen in Equation 2. From the TIM test rigs we get $R_{\text{tot}}^{\text{TIM}}$ seen in Equation 3. We assume that R_{spacers} is the same in both measurements. $R_{\text{air}}^{\text{Ref}}$ is scaled with the smaller interface area to get $R_{\text{air}}^{\text{TIM}}$. Then we can calculate R_{TIM} with the help of Equation 4. The calculated thermal resistances for the 1 mm TIMs are plotted in Fig. 7 for the thermal aging and damp heat tests.

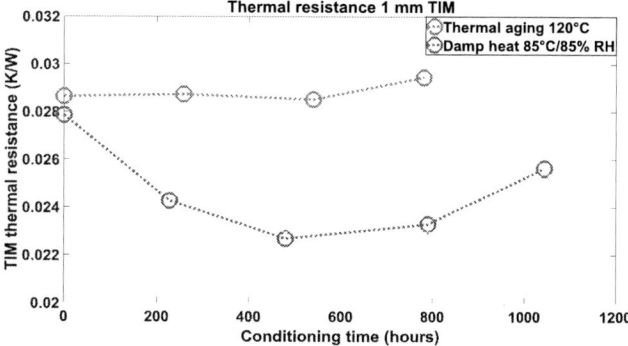

Fig. 7: TIM thermal resistance plotted against conditioning time for the 1 mm TIMs after subtracting the contribution of the test rigs.

In Fig. 7 different trends can be observed for the TIM thermal resistance in the thermal aging and damp heat test. Performance is rather constant over time for the TIM in the thermal aging test. The thermal resistance of the TIM in the damp heat test decreases at first but start to increase at some time after 480 hours. It is worth noticing that the change in performance is very close to what have been observed in Fig. 4 and 5. This means that the contribution from the screws and spacers in the test rigs is small compared to the TIM itself. This is anticipated because the thermal resistances of the reference rigs are generally about one order of magnitude higher than the TIM test rigs. An analysis of Equation 3 reveals that if R_{spacers} and $R_{\text{air}}^{\text{TIM}} \gg R_{\text{TIM}}$ then $R_{\text{TIM}} \approx R_{\text{tot}}^{\text{TIM}}$.

IV. Conclusion

The reliability tests are ongoing and will continue until a minimum of 1 000 hours, TIM failure or no more observable degradation.

So far, the tested graphene enhanced thermal pads have stable thermal performance over time. However, several trends could be observed for the different reliability tests.

The reference rigs show a general trend of decreasing thermal resistance over time. This entails a slight contribution on the total thermal interface resistance for the TIM test rigs as well.

In the thermal aging test, the thermal resistance of the 0.8 mm and 1 mm TIM is rather constant over the test time. The 2 mm TIM shows an initial increase in thermal interface resistance of 16% after 540 hours. In the measurement at 780 hours the increase is only 6%, this is possibly attributed to a longer time in room temperature before the steady state measurement was carried out. In the damp heat test, the thermal interface resistance initially decreased for all the TIM specimens. The trend shifted some time after 480 hours and the total thermal interface resistance started to increase instead. In the thermal cycling test the maximum increase in thermal interface resistance so far was measured to be 13% for the 0.8 mm TIM after 511 cycles.

It was possible to eliminate the contribution of the test rigs for the 1 mm TIM specimens. This was done by subtracting the thermal resistance of the reference rigs from the test rigs containing the TIMs. The result after eliminating the contribution from the test rig is similar to the total thermal interface resistance. This is due to that the thermal resistance of the reference rigs are generally about one order of magnitude higher than the TIM itself.

The test method is deemed to be promising for future reliability studies of thermal pads. If the rig design and the thermal characterization parameters are standardized, then the method can be used for reliability comparison of thermal pads.

Acknowledgment

M. Enmark and J. Liu acknowledge the funding provided by 2D-TECH Vinnova competence center (Ref. 2019-00068). In addition, Y. Fu and J. Liu also acknowledge the financial support from the Swedish National Science Foundation with the contract No: 621-2007-4660 and from the Production Area of Advance at Chalmers University of Technology.

References

[1] J. Hansson, T. M. J. Nilsson, L. Ye, and J. Liu, "Novel nanostructured thermal interface materials: a review," *International Materials Reviews*, vol. 63, no. 1, pp. 22–45, 2018.

[2] Y. Fu, J. Hansson, Y. Liu, S. Chen, A. Zehri, M. K. Samani, N. Wang, Y. Ni, Y. Zhang, Z.-B. Zhang, *et al.*, "Graphene related materials for thermal management," *2D Materials*, vol. 7, no. 1, p. 012001, 2019.

[3] A. A. Balandin, S. Ghosh, W. Bao, I. Calizo, D. Teweldebrhan, F. Miao, and C. N. Lau, "Superior thermal conductivity of single-layer graphene," *Nano letters*, vol. 8, no. 3, pp. 902–907, 2008.

[4] D. Nika, S. Ghosh, E. Pokatilov, and A. Balandin, "Lattice thermal conductivity of graphene flakes: Comparison with bulk graphite," *Applied Physics Letters*, vol. 94, no. 20, p. 203103, 2009.

[5] D. Ghosh, I. Calizo, D. Teweldebrhan, E. P. Pokatilov, D. L. Nika, A. A. Balandin, W. Bao, F. Miao, and C. N. Lau, "Extremely high thermal conductivity of graphene: Prospects for thermal management applications in nanoelectronic circuits," *Applied Physics Letters*, vol. 92, no. 15, p. 151911, 2008.

[6] P. Kim, L. Shi, A. Majumdar, and P. L. McEuen, "Thermal transport measurements of individual multiwalled nanotubes," *Physical review letters*, vol. 87, no. 21, p. 215502, 2001.

[7] E. Pop, D. Mann, Q. Wang, K. Goodson, and H. Dai, "Thermal conductance of an individual single-wall carbon nanotube above room temperature," *Nano letters*, vol. 6, no. 1, pp. 96–100, 2006.

[8] F. Sarvar, D. C. Whalley, and P. P. Conway, "Thermal interface materials-a review of the state of the art," in *2006 1st electronic systemintegration technology conference*, vol. 2, pp. 1292–1302, IEEE, 2006.

[9] X. Guo, S. Cheng, W. Cai, Y. Zhang, and X.-a. Zhang, "A review of carbon-based thermal interface materials: Mechanism, thermal measurements and thermal properties," *Materials & Design*, vol. 209, p. 109936, 2021.

[10] Q. Liang, X. Yao, W. Wang, Y. Liu, and C. P. Wong, "A three-dimensional vertically aligned functionalized multilayer graphene architecture: an approach for graphene-based thermal interfacial materials," *ACS nano*, vol. 5, no. 3, pp. 2392–2401, 2011.

[11] Y.-F. Zhang, D. Han, Y.-H. Zhao, and S.-L. Bai, "High-performance thermal interface materials consisting of vertically aligned graphene film and polymer," *Carbon*, vol. 109, pp. 552–557, 2016.

[12] N. Wang, S. Chen, A. Nkansah, Q. Wang, X. Wang, M. Chen, L. Ye, and J. Liu, "Vertically aligned graphene-based thermal interface material with high thermal conductivity," in *2018 24rd International Workshop on Thermal Investigations of ICs and Systems (THERMINIC)*, pp. 1–4, IEEE, 2018.

[13] N. Wang, S. Chen, A. Nkansah, L. Ye, and J. Liu, "Light-weight compressible and highly thermal conductive graphene-based thermal interface material," in *2018 7th Electronic System-Integration Technology Conference (ESTC)*, pp. 1–5, IEEE, 2018.

[14] "Standard test method for thermal diffusivity by the flash method," tech. rep., ASTM International, West Conshohocken, PA, Apr. 2022.

[15] "Standard test method for thermal transmission properties of thermally conductive electrical insulation materials," tech. rep., ASTM International, West Conshohocken, PA, Nov. 2017.

[16] J. Due and A. J. Robinson, "Reliability of thermal interface materials: A review," *Applied Thermal Engineering*, vol. 50, no. 1, pp. 455–463, 2013.

[17] "Steady-state temperature-humidity bias life test," tech. rep., JEDEC Solid State Technology Association 2021, Arlington, VA, Jan. 2021.

[18] "Temperature cycling," tech. rep., JEDEC Solid State Technology Association 2020, Arlington, VA, Nov. 2020.

Soft, Stretchable and Wireless Sensor Patch with Digitally Printed Liquid Metal Alloy Interconnects

Jan Maslik*, Oskar Hellman*, Bei Wang*, Alessandro Gumiero†, Luigi Dellatorre†, Gustaf Mårtensson‡ and Klas Hjort*

*Division of Microsystems Technology, Department of Materials Science and Engineering
Ångström Laboratory, Uppsala University, 751 03 Uppsala, Sweden
Email: jan.maslik@angstrom.uu.se
†STMicroelectronics, 208 64 Agrate Brianza, MB, Italy
‡Mycronic AB, 183 03 Täby, Sweden

Abstract— Characteristics of high electrical conductivity, high strain tolerance and resistance to fatigue are vital for electronic circuits of on-skin wearable systems. Gallium-based liquid metals offer a unique combination of these characteristics making them excellent alternatives to conventional conductive stretchable inks. In order to obtain better wearing experience, it is advantageous to fabricate devices using breathable materials. However, effective automation solutions for the production of high-resolution digitally patterned circuits for soft and stretchable devices remain a challenge. The presented manufacturing strategy involves adopting a needle dispensing technique for the precise patterning of liquid metal conductors. The circuitry is deposited onto a soft, thin and highly breathable polyurethane medical film. Further, we investigate and map conditions of reliable printing of liquid metal on the polyurethane film for two sizes of dispensing needles with inner diameters of 150 μm and 360 μm. Despite the increased porosity and surface roughness associated with the high breathability of the film, it is possible to reliably deposit liquid metal interconnects with a line width and height below 100 μm. The technological solution results in a first demonstrator presented: an electrophysiological patch.

Keywords—wearable electronics, soft electronics, digital printing, liquid metal patterning

I. INTRODUCTION

Monitoring of biometrics is a rapidly growing field in both sports, healthcare, as well as in everyday life. Providing high compliance and comfort, soft and stretchable wireless sensors and smart patches open up for new applications in the realm of wearable electronics [1 - 4].

The presented wearable device, consisting of a disposable patch and reusable electronic modules, acquires signals of the electrical activity of the heart, wirelessly streams data to a computer, and displays a real-time electrocardiogram. To link the two parts of the device, the patch carries a small and thin flexible circuit with a mounted rigid connector. The assembly of the device takes place on a large area stretchable carrier, approximately an A4 size, and is performed by automated pick-n-place technology. Digitally printed liquid metal (LM) traces serve as compliant electrical interconnects between the epidermal ECG electrodes and the flexible circuit with the rigid connector.

The reusable modules are of two kinds. The first module contains a sensor and data collecting circuit board with Bluetooth communication capabilities. The second module consists of a rechargeable cell battery that functions as a power supply for the sensor module. Both the sensor board and the battery are mounted on a thin flexible board with a rigid connector through which the modules can be coupled together and with the patch. The whole rigid-flex assembly is embedded in a soft silicone gel with a low-friction polyurethane coating contributing to the unobtrusiveness of the device for a wearer.

The most common method to fabricate stretchable electronics is to use meandering spring structures of thin solid conductors. Circuits of this kind are often limited to low elongations with out-of-plane deformation due to the yield strain of the conductor. As a consequence, contact fatigue of integrated rigid components yield from cyclic strain [5 - 6]. In addition, these thin meandering interconnects need a larger footprint and have higher electrical resistance than those in normal rigid and flexible circuits.

Allowing for straight lines and large cross sections with high compliance, LM conductors of gallium-based alloys, such as E-GaIn or Galinstan, show excellent electrical and thermal characteristics, which makes them good candidates for use in soft and compliant electronics for biometric sensing. However, circuits of LM are challenging to manufacture with conventional PCB technologies and therefore the focus of this study is on the digital printing of the LM for the disposable patch. Various deposition technologies and methods have in the recent years been studied and the patterning techniques can be divided into two main groups, i.e., masked deposition [7 - 8] and digital printing. Digital printing has been demonstrated by ink jet printing and needle dispensing [9 - 12].

As previously studied, the deposition of liquid metals through needle dispensing is driven by a shearing mechanism rather than a volumetric or pressure displacement mechanism typical for viscoelastic liquids. Parameters like print height and applied extruding pressure are crucial for quality and reliable patterning. Galinstan naturally forms a nanometer thin oxide layer when in contact with air, and the hydrophilic nature of this oxide contacting the surface allows for adhesion of the liquid metal to many substrates during deposition [13]. The choice of substrate material and its surface conditions are therefore also of importance for reliable and consistent deposition. The parameters of acceleration and deposition speed have no significant effect on the geometry of the patterned structures and therefore no attention is paid to it [14 - 15].

II. EXPERIMENTAL

A. Substrate preparation

Substrates for patterning were prepared by spin coating of approximately 100 μm films of PDMS Sylgard 182 (DOW, Michigan, USA) on 10 × 10 cm glass panes. Thermoplastic polyurethane film, Inspire®2150 (TC Transcontinental, Montréal, Canada) cut to size and removed from its paper liner was transferred to the PDMS coated glass to electrostatically adhere.

B. Liquid metal patterning

Galinstan (Geratherm, Germany) alloy consisting of approximately 68% gallium, 22% indium, and 10% tin, with smaller amounts of Bi and/or Sb, was used for the patterning of LM interconnects. LM patterning was realized with a customized setup consisting of a table top robot, MYT50 (Mycronic, Täby, Sweden); pressure applied to a syringe cartridge was controlled by a precision dispenser Ultimus V (Nordson EFD, Ohio, USA); and the process was monitored by a digital microscope (Dino-Lite Europe, The Netherlands). Patterning was performed using two needles of different inner diameter, a custom-made glass capillary needle of inner diameter 150 μm (Polymicro Technologies/Molex, Phoenix, USA), and an Optimum® flexible dispensing tip (Nordson EFD, Ohio, USA) of inner diameter 360 μm.

In order to identify parameters within which continuous and uniform traces are dispensed a preliminary study on the stable pressure-print height relation was performed for both needles. Deposition speed was set to 1 mm/s for patterning. Liquid metal traces were printed at a set printing height (i.e. proximity of printing needle from the substrate surface), while increasing the pressure in the cartridge up to the point where the oxide layer yielded, forming droplets on the substrate. The procedure was repeated increasing the printing height in increments of 10 μm until the LM trace lost contact with the substrate. The inspection was performed three times for each of the dispensing needles. For a detailed map of the parameters for reliable dispensing, refer to Fig. 1 and Fig. 2. Based on the preliminary study and identified parameters of reliable dispensing a print repeatability test of five examination lines was subsequently performed. Characterization of the influence of print height and applied pressure on the geometry of deposited traces was evaluated.

The repeatability test was done by printing 3 cm traces at specific printing conditions across the stable pressure-print height area. The width and height of the traces were measured at three different spots along each trace with vertical scan interferometry (3D Optical Profiler ZYGO, AMETEK, Berwyn, USA) using 10x magnification. The values of width and height of the traces were averaged and supplemented by a deviation.

C. Patch fabrication

The devices were fabricated in batches of six on A4-sized substrates of thermoplastic polyurethane film supplied on a paper liner Inspire®2150 (TC Transcontinental, Wrexham, United Kingdom). Using a spray gun, highly conductive paste (124-36, Creative Materials, Ayer, USA) was after thinning sprayed, transferring epidermal electrode patterns on top of the film. The original formulation of the paste contains 80 wt% Ag/AgCl particles with ratio of 66:34. The consistency of the highly viscous paste was modified by adding a designated thinner (102-03, Creative Materials, Ayer, USA) in 1:1 weight ratio. Molex 0.5 mm Pitch SlimStack Plug connectors (555600207) soldered to copper clad polyimide laminate (Holders Technology, London, UK) with contact pads for the LM, were cut to size and glued to the polyurethane substrate using a silicone adhesive (Elastosil A07, Wacker Chemie, Munich, Germany). Positioning of the components was performed using an automated pick-and-place robot MY300DX (Mycronic AB, Sweden). Patterning of LM interconnects between the connector and the electrodes was achieved as described above using the Optimum® flexible dispensing tip (Nordson EFD, Ohio, USA) of inner diameter 360 μm. Patterning was done with 1.3 kPa of pressure and 60 μm print height. The encapsulation of the LM interconnects was performed using the same setup as for patterning LM, but using the silicone adhesive (Elastosil A07, Wacker Chemie, Munich, Germany). The print height was set at 1 mm and velocity 10 mm/s. A smooth flow tapered dispensing tip G22 (ID of 0.41 mm) (Fisnar, Germantown, USA) was used as a printing needle. The applied pneumatic pressure was set to 120 kPa. The material used as a skin adhesive is a medical grade soft silicone gel adhesive, Silbione 4645 (ELKEM Silicones, Oslo, Norway), that was selectively dispensed in the form of fringes around the circumference of the patch to maintain high breathability. The print height was set at 0.3 mm and velocity 10 mm/s. Optimum® SmoothFlow™ tapered dispensing tip G25 with an inner diameter of 250 μm (Nordson EFD, Ohio, USA) was used as a printing needle. The applied pneumatic pressure was set to 250 kPa. In order to establish a stabile electrical interface between the epidermal electrode and the skin, sensing hydrogel pads (AG625, Axelgaard Manufacturing, Fallbrook, USA) were attached to the Ag/AgCl electrodes.

D. Integration of electronics

A prototype of a sensor device for data acquisition and streaming via Bluetooth (ST Microelectronics, Italy) was reconfigured in a new, soft casing with a counter connector Molex 0.5 mm Pitch SlimStack PCB Receptacle (541020204) to fit the patch. To power the sensor device, a rechargeable lithium coin cell battery LIR2032 was used and cased in the same fashion.

III. RESULTS AND DISCUSSION

A. Liquid metal patterning

The constructed map of parameters serves as a springboard for defining specific process window of continuous and uniform dispensing and for the subsequent characterization of the influence of these parameters on the geometry of deposited traces.

The regions of dispensing modes mapped in Fig. 1 and Fig. 2, respectively, can be characterized as follows: a region of discontinuous dispensing refers to too low proximity of the needle to the substrate – results in no dispensing or dispensing with poor quality affecting conductive electrical performance. A region of low feeding pressure implies imperfect dispensing caused by too low pressure applied in a cartridge supplying LM adequately during dispensing process. In contrast, too high pressure applied in the cartridge causes the deposited LM channel held by the oxide shell to be overfilled. Therefore, the oxide skin yields and the dispensing process results in the uncontrolled formation of blobs of LM. If the distance between the printing needle and the substrate is too high, there is a loss of contact between the LM at the orifice of a needle and the substrate surface. The LM cannot be further sheared on top of the surface and the mode leads to necking along the deposited lines or completely losing contact resulting in dispensing of disjointed traces.

The parameters at which it was possible to deposit continuous traces and which overlapped in the inspections were determined as the region of stable dispensing parameters. These regions are marked with the black frame.

In the case of the 150 μm ID needle, the stable printing region was defined to be between 30 and 60 μm of print height with 6 to 8 kPa of applied pressure. Furthermore, the influence of these parameters on the dimensions of printed lines deposited on a breathable polyurethane substrate was investigated, similar to previous studies by Cook et al. [14]. First of all, a constant print height of 45 μm with variable pressure of 6, 7 and 8 kPa. An analysis of dimensions of patterned traces as a function of pressure is plotted in Fig. 3 with an image of LM traces in Fig. 4. Further, a constant pressure of 7 kPa was determined with varying printing heights of 30, 40, 50 and 60 μm. Analysis of traces as a function of print height are plotted in Fig. 5 and depicted in Fig. 6.

The parameters of printing height and the applied pressure independently affect the geometry of the deposited structures. However, it is possible to observe a certain inconsistency in the geometry of the lines along their lengths, as evidenced by the standard deviation. This fact can be associated both with the deviation of the calibration of the exact print height, but also with LM wetting of the rugged surface topography of the polyurethane film. For surface analysis of used polyurethane film, see Fig. 7. However, the applied pressure significantly affects, both the dispensing itself and the finishing phase of printing. An appropriate combination of switching off the applied pressure with a sharp lift-off of the print head allows for the formation of smooth lines.

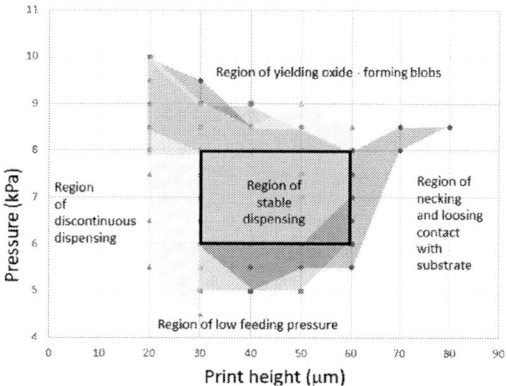

Fig. 1. Map of dispensing modes using 150 μm inner diameter printing needle with a region of stable dispensing marked with black frame.

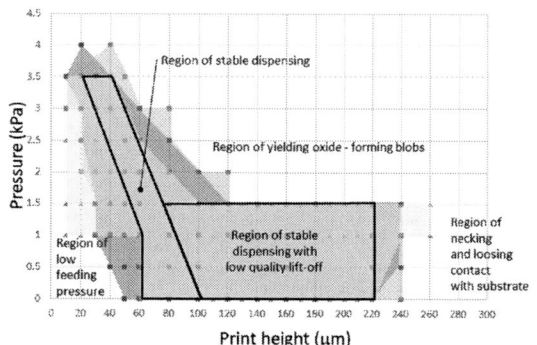

Fig. 2. Map of dispensing modes using 360 μm inner diameter printing needle with a region of stable dispensing marked with black frame.

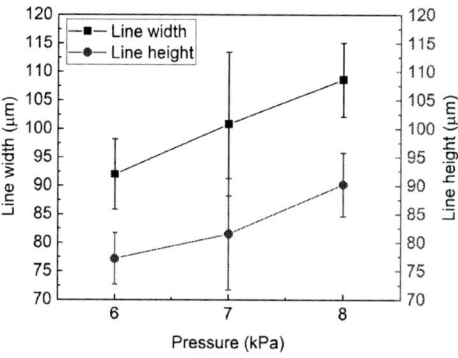

Fig. 3. Trace dimensions analysis of LM printed with 150 μm needle at constant print height of 45 μm and varying pressure.

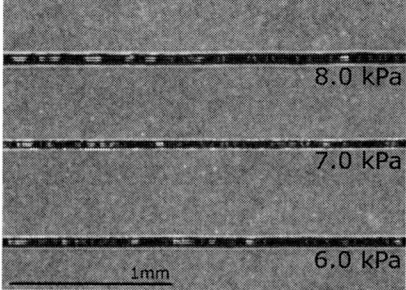

Fig. 4. Image of LM traces printed with 150 μm needle at constant print height of 45 μm and varying pressure.

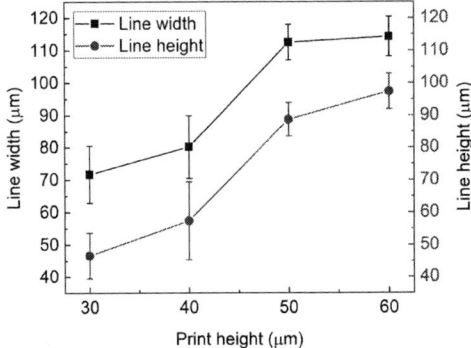

Fig. 5. Trace dimensions analysis of LM printed with 150 μm needle at constant pressure of 7 kPa and varying print heights

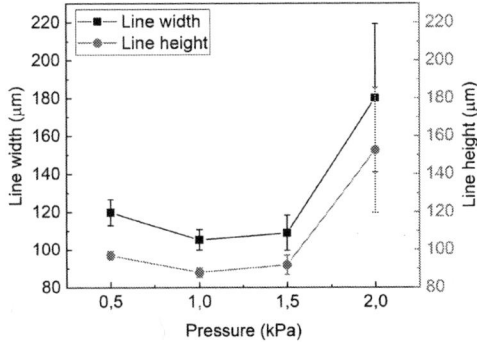

Fig. 8. Trace dimensions analysis of LM printed with 360 μm needle at constant print height of 60 μm and varying pressure.

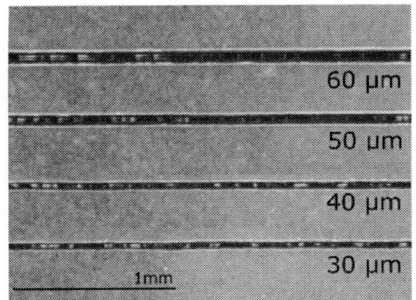

Fig. 6. Image of LM printed with 150 μm needle at constant pressure of 7 kPa and varying print heights

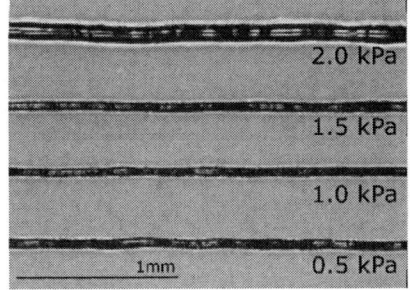

Fig. 9 Image of LM traces printed with 360 μm needle at constant print height of 60 μm and varying pressure

A similar analysis was performed using a printing needle with an inner diameter of 360 μm. Based on the mapped region of reliable dispensing, a constant print height of 60 μm was determined for the reproducibility test with varying applied pressure of 0.5, 1, 1.5 and 2 kPa. The analysis of the dimensions of the patterned traces as a function of pressure are plotted in Fig. 8 and Fig. 9. Analysis of traces as a function of print height are plotted and depicted in Fig. 10 and Fig. 11.

As previously reported [12, 14] and confirmed in our analysis, it is possible to reliably deposit and pattern LM traces using larger inner diameter printing needles at higher print heights, making the patterning technique less sensitive to eventual substrate unevenness. This is very important for large-scale deposition areas, A4 size as an example, for potential industrial adaptation. Our experience indicates a certain level of

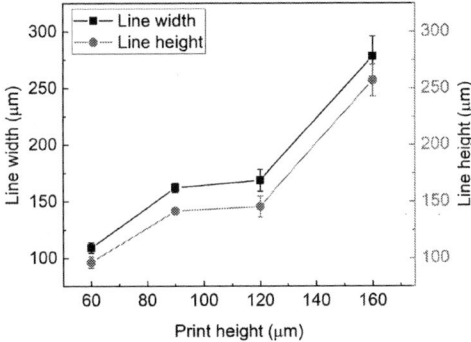

Fig. 10. Trace dimensions analysis of LM traces printed with 360 μm needle printed at constant pressure of 1 kPa and varying print heights

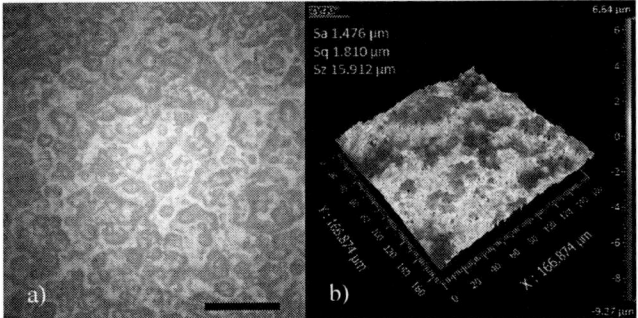

Fig. 7. Surface analysis of used polyurethane film (Inspire®2150, TC Transcontinental). a) Micrograph of surface and b) 3D topography of scanned surface. Scale bar corresponds to 50 μm.

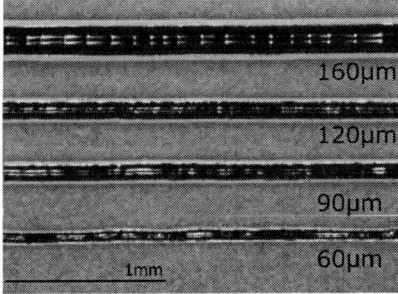

Fig. 11. Image of LM traces printed with 360 μm needle printed at constant pressure of 1 kPa and varying print heights.

sensitivity to pressure control of feeding pressure and places higher demands on the precision of pressure control. The waviness/corrugation along the deposited traces is observable especially on traces printed with the 360 μm needle and to some extent on traces printed with the 150 μm needle. According to our data, the waviness occurs when the printing height is less than one third of the inner diameter of the needle. This phenomenon may have a more complex cause, but indicates a more significant effect of surface roughness at lower printing heights. When printing with the 150 μm inner diameter needle, the geometry of the traces are highly dependent on print height. A lower print height results in traces with high width/height ratio significantly, see Fig. 12. It should also be noted that the deviation from the mean width/height ratio is larger for the traces printed at lower heights due to the relative change of height related to unevenness of the surface. This deviation relates to the shearing mechanism of LM on top on surface with increased surface roughness. The surface of increased roughness results in inconsistent wetting by LM, and thereby an uneven trace width [16 - 17]. We assume that for ultra-high-resolution dispensing using needles with lower inner diameters, surface topography and associated roughness can have significant impact. Depositing LM at higher print heights, both with the 150 μm needle and the 360 μm needles, results in lower width/height ratio and smaller deviations from the mean, Fig. 12 and Fig. 13, respectively. This indicates a mechanism in which the interface

between the substrate surface and the sheared LM is smaller and therefore LM forms a stable tubular filament-like geometry.

B. Wearable patch functioning

The functionality of the fabricated wearable sensor device consisting of the wearable patch with LM interconnects and modular electronics depicted in the Fig. 14 was tested by connecting the modules with the patch and first applied on the forearm skin of a male wearer, Fig. 15. Wearing aspects, such as adhesion to the skin, suitable pressing of the electronic modules to the skin, and overall unobtrusiveness under the clothing over a longer wearing period, were taken into consideration. Furthermore, the aspect of wireless connection, data acquisition and transmission were verified.

For real-time electrocardiography (ECG) monitoring, the wearable sensor device was applied in the middle of the top of the abdomen of a male wearer as illustrated in Fig. 16 with a section of the electrocardiogram collected from the wearable patch.

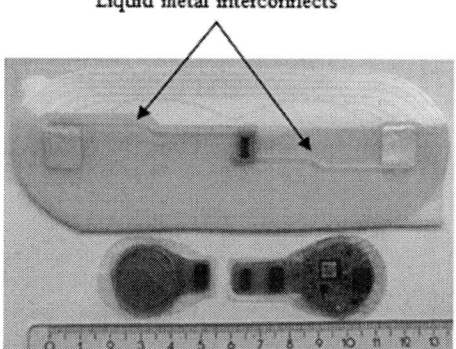

Fig. 14. A photograph of the wireless wearable device consisting of electronic modules and soft and stretchable sensor patch with patterned liquid metal interconnects.

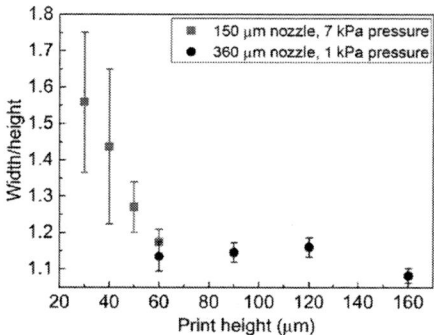

Fig. 12. Width/height aspect ratio relation for the printed traces at constant pressure

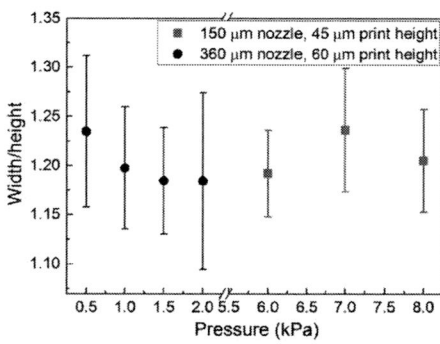

Fig. 13. Width/height aspect ratio relation for the printed traces at constant print height.

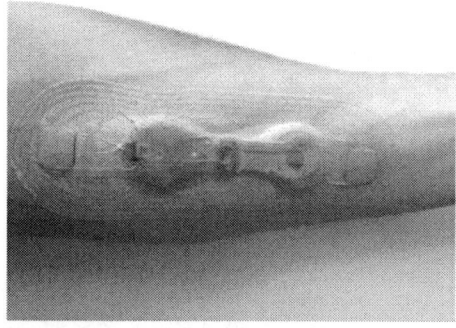

Fig. 15. A photograph of attached wireless sensor patch on forearm skin for demonstration.

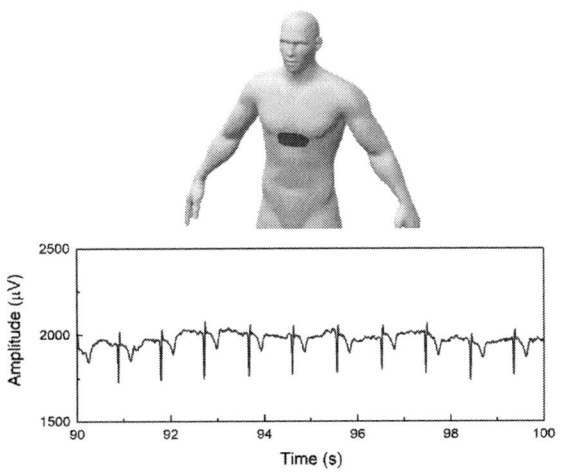

Fig. 16. Positioning of the wearable sensor patch in the middle of the top of the abdomen marked as a blue area with an example of monitored electrocardiogram.

IV. CONCLUSION

We inspected and demonstrated LM patterning technique on a non-silicone elastomeric substrate with considerable breathability. The investigation of dispensing parameters for 150 μm and 360 μm inner diameter needles provides a reference to reliably pattern LM on a porous and rough surface. However, the precision and repeatability of patterning are subject to previously reported conditions such as surface uniformity, evenness and cleanliness. To overcome surface unevenness across larger deposition areas, dispensing needles of higher inner diameters offer a simple solution for potential scaled-up manufacturing. The digitally patterned LM traces were incorporated as compliant electrical interconnects in the design of the wearable device. The fabricated soft, stretchable and breathable sensor patch exhibited the ability to monitor electrocardiography and wirelessly stream the signal in real-time which proved the high potential for applications in sport and healthcare.

ACKNOWLEDGMENT

The authors would like to thank Klara Björnander Rahimi for her contribution in the initial stage of the work, Elvis Carlsson for guidance and assistance with pick and place technology, and TC Transcontinental UK for providing sample materials.

This research was funded by the European Union's Horizon 2020 research and innovation programme under grant agreement No 824984 - Soft intelligence epidermal communication platform (SINTEC).

REFERENCES

[1] S. Xu, A. Jayaraman, and J.A. Rogers, "Skin sensors are the future of health care," Nature, vol. 571(7765), pp. 319-321, July 2019.

[2] K.Y. Kwon, Y.J. Shin, J.H. Shin, et al., "Stretchable, Patch-Type Calorie-Expenditure Measurement Device Based on Pop-Up Shaped Nanoscale Crack-Based Sensor," Adv. Healthc. Mater., vol. 8(19), October 2019.

[3] H.U. Chung, B.H. Kim, J.Y. Lee, et al., "Binodal, wireless epidermal electronic systems with in-sensor analytics for neonatal intensive care," Science, vol. 363(6430) eaau0780, March 2019.

[4] J. Alberto, C. Leal, C. Fernandes, et al., "Fully Untethered Battery-free Biomonitoring Electronic Tattoo with Wireless Energy Harvesting," Sci. Rep., vol. 10, pp. 5539, March 2020.

[5] K.D. Harris, A.L. Elias, and H.-J. Chung, "Flexible electronics under strain: a review of mechanical characterization and durability enhancement strategies," J. Mater. Sci., vol. 51, pp. 2771–2805, March 2016.

[6] K.-I. Jang, K. Li, H. Chung, et al., "Self-assembled threedimensional network designs for soft electronics," Nat. Commun. vol. 8, pp. 15894, June 2017.

[7] S. Jeong, K. Hjort, and Z. Wu, "Tape Transfer Atomization Patterning of Liquid Alloys for Microfluidic Stretchable Wireless Power Transfer," Sci. Rep., vol. 5, pp. 8419, February 2015.

[8] B. Wang, J. Gao, J. Jiang, Z. Hu, K. Hjort, Z. Guo, and Z. G. Wu, "Liquid Metal Microscale Deposition enabled High Resolution and Density Epidermal Microheater for Localized Ectopic Expression in Drosophila," Adv. Mater. Technol., vol. 7, pp. 2100903, March 2022.

[9] G. Li, X. Wu and D.-W. Lee, "A galinstan-based inkjet printing system for highly stretchable electronics with self-healing capability," Lab on a chip, vol. 16(8), pp. 1366-1373, March 2016.

[10] J.W. Boley, E.L. White, G.T.-C Chiu, and R.K. Kramer, "Direct Writing of Gallium-Indium Alloy for Stretchable Electronics," Adv Funct. Mater., vol. 24, pp. 3501-3507, February 2014.

[11] Y. Yoon, S. Kim, D. Kim, S.K. Kauh, J. Lee, "Four Degrees-of-Freedom Direct Writing of Liquid Metal Patterns on Uneven Surfaces," Adv. Mater. Technol., vol. 4(2), pp. 1800379, October 2018.

[12] S. Kim, J. Oh, D. Jeong, W. Park, and J. Bae, "Consistent and Reproducible Direct Ink Writing of Eutectic Gallium–Indium for High-Quality Soft Sensors," Soft Robotics, vol. 5(5), pp. 601-612, October 2018.

[13] K. Doudrick, S. Liu, E. M. Mutunga, K. L. Klein, V. Damle, K. K. Varanasi, and Konrad Rykaczewski, "Different Shades of Oxide: From Nanoscale Wetting Mechanisms to Contact Printing of Gallium-Based Liquid Metals," Langmuir, vol. 30(23), pp. 6867-6877, May 2014.

[14] A. Cook, D.P. Parekh, C. Ladd, G. Kotwal, L. Panich, M. Durstock, M.D. Dickey and C.E. Tabor, "Shear-Driven Direct-Write Printing of Room-Temperature Gallium-Based Liquid Metal Alloys," Adv. Eng. Mater., vol. 21(11), pp. 1900400, July 2019.

[15] M.D. Dickey, R.C. Chiechi, R.J. Larsen, E.A. Weiss, D.A. Weitz, and G.M. Whitesides, "Eutectic Gallium-Indium (EGaIn): A Liquid Metal Alloy for the Formation of Stable Structures in Microchannels at Room Temperature," Adv. Funct. Mater., vol. 18, pp. 1097-1104, 2008.

[16] R.K. Kramer, J.W. Boley, H.A. Stone, J.C. Weaver, and R.J. Wood, "Effect of microtextured surface topography on the wetting behavior of eutectic gallium-indium alloys," Langmuir, vol. 30 (2), pp. 533-539, January 2014.

[17] I.D. Joshipura, H. R. Ayers, G. A. Castillo, C. Ladd, C. E. Tabor, J. J. Adams, and M. D. Dickey, "Patterning and Reversible Actuation of Liquid Gallium Alloys by Preventing Adhesion on Rough Surfaces," ACS Applied Materials & Interfaces, vol. 10 (51), pp. 44686-44695, 2018.

IEEE
445 Hoes Lane
Piscataway, NJ 08854-4141

ISBN 978-1-6654-9177-8